WORLD SOYBEAN RESEARCH CONFERENCE II: ABSTRACTS

Other Titles of Interest

New Agricultural Crops, edited by Gary A. Ritchie

Successful Seed Programs: A Planning and Management Guide, Johnson E. Douglas

Animals, Feed, Food and People: An Analysis of the Role of Animals in Food Production, edited by R. L. Baldwin

Pest Control: Cultural and Environmental Aspects, edited by David Pimentel and John H. Perkins

Pest Management in Transition: With a Regional Focus on the Interior West, Pieter de Jong, project coordinator

Small-Scale Processing and Storage of Tropical Root Crops, edited by Donald L. Plucknett

The Mineralogy, Chemistry, and Physics of Tropical Soils with Variable Charge Colloids, Goro Uehara and Gavin P. Gillman

Azolla as an Aquatic Green Manure: Use and Management in Crop Production, Thomas A. Lumpkin and Donald L. Plucknett

Irrigated Rice Production Systems: Design Procedures, Jaw-Kai Wang and Ross Ellis Hagan

Managing Pastures and Cattle Under Coconuts, Donald L. Plucknett

Rice in the Tropics: A Guide to the Development of National Programs, Robert F. Chandler, Jr.

About the Book and Editor

World Soybean Research Conference II: Abstracts
edited by Frederick T. Corbin

The result of strong international interest in the soybean, the World Soybean Research Conference II was held March 26–29, 1979, at North Carolina State University. This volume contains summaries of the more than two hundred papers presented at that meeting. The authors, international authorities in their fields, represent sixteen areas of professional specialization. In this overview of major soybean research conducted since 1975, they describe specific accomplishments and advances in entomology, pathology, engineering, economics, food science, genetics, physiology, and agronomy. In addition, various aspects of research on soybean production, marketing, and utilization are examined. Also available from Westview Press is *World Soybean Research Conference II: Proceedings,* which contains seventy-four of the invited papers in full.

Frederick T. Corbin is professor of crop science at North Carolina State University in Raleigh.

Published in cooperation with
North Carolina State University

WORLD SOYBEAN RESEARCH CONFERENCE II
March 26–29, 1979

CONFERENCE SPONSORS

Agency for International Development
American Soybean Association Research Foundation
INTSOY/University of Illinois
National Soybean Crop Improvement Council
North Carolina Agricultural Extension Service
North Carolina Agricultural Research Service
North Carolina Crop Improvement Association
North Carolina Foundation Seed, Inc.
North Carolina Soybean Producers Association
School of Agriculture and Life Sciences, NCSU
Science and Education Administration, USDA

B.E. Caldwell, Conference Chairman

PUBLICATIONS COMMITTEE

F. T. Corbin, Department of Crop Science, NCSU
H. T. Daniel, Department of Economics, NCSU
D. P. Schmitt, Department of Plant Pathology, NCSU
J. W. Van Duyn, Department of Entomology, NCSU
P. A. Wilson, PawPrint Graphics and Typesetting

WORLD SOYBEAN RESEARCH CONFERENCE II:

ABSTRACTS

edited by

Frederick T. Corbin

Routledge
Taylor & Francis Group
NEW YORK AND LONDON

First published 1980 by Westview Press, Inc.

Published 2018 by Routledge
605 Third Avenue, New York, NY 10017
2 Park Square, Milton Park, Abingdon, Oxon OX14 4RN

Routledge is an imprint of the Taylor & Francis Group, an informa business

Library of Congress Cataloging in Publication Data
World Soybean Research Conference, 2d, North Carolina State University, 1979.
Abstracts.
Includes index.
1. Soybean—Congresses. 2. Soybean products—Congresses. I. Corbin, Frederick T.
SB205.S7W67 1979 635'.655 80-10381
ISBN 0-89158-899-X

Published in Great Britain 1980 by
Granada Publishing Limited - Technical Books Division
Frogmore, St Albans, Herts AL2 2NF
and
3 Upper James Street, London W1R 4BP
117 York Street, Sydney, NSW 2000, Australia
PO Box 84165, Greenside, 2034 Johannesburg, South Africa
61 Beach Road, Auckland, New Zealand

British Library Cataloguing in Publication Data
World Soybean Research Conference, *2nd, Raleigh, 1979*
World Soybean Research Conference II, Abstracts
1. Soybean—Congresses
I. Corbin, Frederick T.
633'.34 SB205.S7

ISBN 13: 978-0-367-21404-3 (hbk)

TABLE OF CONTENTS

WORLD SOYBEAN RESEARCH CONFERENCE II: ABSTRACTS

KEYNOTE ADDRESSES

DESIRABLE QUALITIES OF SOYBEANS FOR THE WORLD MARKET—JAPANESE CONSUMERS' VIEWPOINTS

H. Nakamura, Hohnen Oil Co., Ltd., Tokyo, Japan

The characteristics of the soybean market in Japan are explained, and desirable qualities of soybeans for different uses are defined. The Japanese soybean market may be said to be the most diversified market in the world, encompassing not only crushing but also various types of food for human consumption. In this sense the desirable qualities in terms of oil content and protein content and fatty acid and amino acid composition as seen from the Japanese standpoint may be a representative view of soybean consumers in the world. These desirable qualities are expected to suggest directions of soybean research in the future.

SHARING AGRICULTURAL TECHNOLOGY

The Honorable Paul Findley, U.S. House of Representatives, Illinois

America's agricultural colleges and universities hold the key to sharing agricultural technology to increase world food production. But the developing countries must help turn the key in the lock if we are to open the door to an effective famine prevention program throughout the world. Title XII of the Foreign Assistance Act—the Famine Prevention Program—provides a framework for success, but there are obstacles to overcome.

MINERAL NUTRITION

Invited Papers

CATION NUTRITION AND ION BALANCE

J. E. Leggett, Agronomy Department, University of Kentucky, Lexington, Kentucky 40506

The rate of cation accumulation by growing plants is very dependent upon the rate of dry matter production. The ratio of cations accumulated is related to the cation ratios in solution, competition among cations, mobility of the anions, and the N source. In the absence of a stress condition, cation accumulation by *Glycine max* will be dominated primarily by the source of N available to the plants. In this case the N source may influence the growth rate, cation accumulation, and the organic acid level. The objective of this study was to evaluate the influence of N source on cation accumulation and these effects on soybean yields.

Soybeans *[Glycine max* (L.) Merr. cv Fiskeby V] were germinated and grown in the gravel culture system. The culture solution was a modified Hoagland solution with N sources as the variable. The treatments included, NH_4^+, urea, NO_3^- and nodule fixation as the sole source. All plants were grown initially on NO_3 then changed at three weeks to the specific N treatments. Samples were collected for analysis at time of treatment initiation, at R1, R4, and R8 growth stages. The plants were separated into leaves, stems + petioles, and pods when present. Analysis included K, Ca, Mg, NO_3, P, S, and organic acids.

Cation accumulation was maximized for those plants grown on NO_3. In addition the organic acid levels were greatest for NO_3 treated plants and much lower for nodulated plants. Interpretation of results between short term and long term experiments must be exercised with caution. Generally the rates can be similar but the growth rate of the plant can influence total accumulation. Hence, with NO_3 as the N source the growth may be greater than for the other N sources, total cation accumulation greater on a per plant basis, but the cation concentration in the leaf may approach equality for all N treatments. It also follows that the grain yield potential is greater for the larger plants.

RESPONSE OF SOYBEANS TO LIMING ON ACID TROPICAL SOILS

F. Abruna, USDA, SEA/FR, Rio Piedras, Puerto Rico

Toxic levels of Al and/or Mn are often the main cause of low crop yields in the tropics. The effect of soil acidity factors on soybean [*Glycine max* (L.) Merrill] were determined in a series of field experiments on 2 Ultisols and 1 Oxisol. Soybeans responsed strongly to liming on the Ultisols and to a lesser extent on the Oxisol. Soybean yields decreased as percent Al saturation or the Al/base ratio increased or pH decreased. On Corozal clay (Aquic Tropudults) yields increased from 62 to 2081 kg/ha, when Al saturation was decreased from 67 to 3 percent. On Coto soil (Tropeptic Haplorthox) highest yields of 3555 kg/ha were recorded at pH 5.3, 5 percent Al saturation, and an Al/base ratio of .05. However, 71 percent of maximum yield was still obtained at 36 percent Al saturation and pH 4.3.

Calcium content of the soybean leaves increased with decreased soil acidity on all the soils. The N content of the leaves increased with decreasing acidity on Corozal soil (Aquic Tropudults). The Ca/Mn ratio correlated with yields on Humatas (Typic Tropohumults), and Coto (Tropeptic Haplorthox).

Response of soybeans to liming in acid tropical soils appears to be related to Ca uptake and N fixation. On Ultisols, toxic levels of exchangeable Al is the main soil acidity factor affecting yields while on Oxisols soluble Mn seems to be a contributing one.

MINERAL NUTRITION AND NODULATION

D. N. Munns, Department of Land, Air and Water Resources, University of California, Davis, California 95616

The term "nodulation" here embraces all the events and processes leading to effective N_2 fixation in root nodules. Nutritional variables can influence nodulation in two ways, sometimes difficult to distinguish. The variable may affect nodulation directly, with resulting effect on N_2 fixation (and on plant growth if N is limiting). This is normally true for acidity, Al, Ca, Mo, combined N, and perhaps sometimes for Mn, S, and P. Alternatively, the variable may influence the growth of the plant, with a resulting effect on nodulation. This kind of response is typical for P, K, most micronutrients, and probably S, water and CO_2. Either way, there can be a large effect on N_2 fixation.

Most nutritional research involving nodulation has been done with pasture and forage legumes. It indicates much quantitative and even qualitative variation between species in their responses. Generalization and extrapolation is uncertain. Nodulation first requires rhizobial growth in the legume rhizosphere, an environment sometimes inhospitably acid and depleted of immobile nutrients. Soybean rhizobia belong to the generally more acid-tolerant (slow-growing) group, but most of those tested have difficulty growing well below pH 5, and of the acid-tolerant strains, few tolerate realistic (50 μM) concentrations of Al. They do tolerate high Mn (200 μM), and low Ca (50 μM) if Ca + Mg is high enough (about 250 μM). Strangely, there are no data on quantitative requirements of rhizobia for Fe and S, and very few for K and P.

Nodule formation (initiation) is sometimes more sensitive than any other phase of the symbiosis, to acidity, Ca deficiency, and Al toxicity. Interactive effects of Ca and pH on nodule initiation appear to be common. Nodule function requires adequate Mo as an ingredient of nitrogenase. Recent evidence implies that some species in some circumstances need more Ca for nodule functions than for nodule initiation or for growth of the N-fertilized plant. This includes data for soybean, in solution culture. Salt (NaCl) can stop fixation in established nodules of *Glycine wightii*, and in other species. Sulfur deficiency interferes with fixation, mainly because of interference with protein synthesis.

Selection for nutritional tolerance may prove worthwhile. Most work in this direction has been with soil acidity. With some legumes (e.g., cowpea, mung) the host is more tolerant than many of the rhizobial strains. Selection of the latter is indicated. The opposite may hold for soybean. At least with the current run of commercial types, the rhizobia may be more tolerant than the hosts.

Contributed Papers

EVALUATION OF SAMPLING TIME AND LEAF POSITION ON CRITICAL Mn DEFICIENCY LEVEL OF SOYBEANS

K. Ohki, F. C. Boswell, M. B. Parker, L. M. Shuman, and D. O. Wilson, University of Georgia, Department of Agronomy, Georgia Station, Experiment, Georgia 30212 and Coastal Plain Station, Tifton, Georgia 31794

A three-year field study was conducted to evaluate the optimum sampling period and leaf position on critical Mn deficiency levels in blade tissue as related to soybean seed yield. Soybean [*Glycine max* (L.) Merr.] cultivar Ransom was grown on Olustee-Leefield sand with added Mn at rates of 0, 5.6, 11.2, 22.4 and 56 kg/ha as $MnSO_4$. Recently matured blade tissues were samples at the V5, R1, R2, R3 and R6 growth stages for analyses. Blade samples from leaf one (top) down to leaf five (bottom) were taken at the R2 growth stage in years two and three.

As sampling time progressed during the season from V5, R1, R2, R3 to R6 growth stages, the critical Mn levels in recently matured blades for the respective growth stages were 11, 10, 12, 19 and 21 μg Mn/g dry weight. No significant effects on critical Mn deficiency levels were noted in growth stages from V5 to R2, but a significant effect of sampling period was noted between the R2 and R3 growth stages. At growth stages R3 and R6, the critical Mn deficiency levels were the same. Relationships between tissue Mn concentration and seed yields for blades one to five gave critical Mn deficiency levels of 18, 13, 11, 11 and 10 μg Mn/g dry weight, respectively, when sampled at the R2 growth stage. The critical Mn level in blade 1 was greater than blade 2 and blade 2 was greater than blades 3, 4, and 5. There was no difference in critical Mn levels among blades 3, 4, and 5. It was concluded that sampling blade 2 at the R2 growth stage represents the appropriate sample for estimating the critical Mn deficiency level as related to soybean yields.

ALLEVIATION OF MANGANESE TOXICITY AT HIGH TEMPERATURES

D. P. Heenan, L. C. Campbell and O. G. Carter, Department of Agronomy and Horticultural Science, University of Sydney, N.S.W. Australia 2006

Glasshouse and field observations have implicated temperature in the development of manganese toxicity symptoms in some species. Under glasshouse conditions, this factor is often not controlled closely. During a study of the differential tolerance of soybean cultivars to high manganese levels the influence of temperature on manganese nutrition was evaluated.

Two soybean cultivars (Lee and Bragg) were grown in solution culture over a range of manganese levels (0.1, 5 and 15 ppm). An initial experiment was conducted in glasshouse chambers at day-night temperatures of 21-18 C and 33-28 C while a second studied the uptake and distribution of manganese over a range of temperatures (20/15, 24/19, 28/23, 33/28 C in growth cabinets.

Increasing the manganese supply from 0.1 to 15 ppm induced toxicity symptoms (crinkle leaf) and reduced dry matter yields at low temperatures. Symptoms of toxicity and yield reductions were more severe for Bragg than Lee. Increasing the temperature reduced the intensity of symptoms shown for both cultivars. High temperatures depressed the reductions in dry matter at high manganese levels resulting in a significant interaction between temperature and manganese supply.

Although Mn concentrations in the total leaf were increased by increasing temperature there was a small growth dilution effect on manganese in young expanding leaves between 20/15 and 28/23 C. The reductions however were not sufficient to explain decreased sensitivity to high manganese supply. A further increase in temperature from 28/23 to 33/28 C increased manganese concentration in the young expanding leaf but reduced intensity of toxicity symptoms. This suggests that the alleviation of manganese toxicity at high temperatures is not the result of a growth dilution of manganese in growing tissue.

DRIS SYSTEM EVALUATION OF SOYBEAN N, P AND K STATUS INFLUENCED BY P-TREATMENT, VARIETY AND SAMPLING STAGE

R. G. Hanson and J. L. Sebaugh, University of Missouri, Columbia, Missouri, and C. M. Borkert, EMBRAPA-CNPSoja, Londrina-PR, Brazil

Low available P is a soil characteristic of the soybean region of southern Brazil. Soil test calibration research was undertaken to determine P requirements. Leaf samples collected at R-2, R-5 and R-7 growth stages and seed samples were analyzed for N, P and K. Three (3) P sources, 2 varieties and 2 locations were used in the study. DRIS systems norms of M. E. Sumner are used to evaluate N, P and K status of the plant.

The first study was completed using the variety Bragg and TSP as the P-source. The yield correlated to P-treatment at r = 0.69**. Correlation between yield and leaf-N was r = 0.80**, 0.77** and 0.84** at stages R-2, R-5, and R-7, respectively. The yield to leaf-P correlation was r = 0.74**, 0.83** and 0.84** for stages R-2, R-5 and R-7, respectively. The yield correlation to leaf-K was r = 0.38* , r = 38* and -0.55**, respectively, for R-2, R-5 and R-7. The DRIS evaluation of N, P, K at zero P-treatment indicated corrective needs of P>N>K at all three stages. With P-fertilization this sequence changed from P>N>K at stages R-2 and R-5 and to K>P>N at R-7.

The second observations included three (3) P sources, (TSP and 2 rock phosphates) variety vicoja. The yield was correlated to P-treatments r = 0.50**. There was little difference in response to P-source. The yield correlation to leaf-N was r = 32*, 0.63** and NS at stages R-2, R-5 and R-7, respectively. The yield correlation to leaf-P was r = 0.57**, r = 0.70** and r = 0.61** at R-2, R-5 and R-7, respectively. The yield was negatively correlated with leaf-K, r = -0.48**, r = -0.42* and r = -51**, respectively at R-2, R-5 and R-7. The DRIS evaluation at zero P-treatment indicated corrective sequence of P>N>K at the three (3) growth stages. With P-correction they changed from P>N>K at stages R-2 and R-5 but to K>P>N at stage R-7.

Seed-N tended not to correlate to leaf-N, while seed-P was highly correlated to leaf-P and seed-K correlated negatively to leaf-K. Leaf N, P and K decreased with maturity and was more pronounced with K. At high yields the DRIS systems did not indicate the same order or corrective needs near maturity (R-7) as at bloom (R-2).

DIAGNOSIS OF NUTRITIONAL DISORDERS IN SOYBEAN BY INORGANIC TISSUE ANALYSIS

N. S. Murali and J. M. Nielsen, Asian Institute of Technology, Bangkok, Thailand and The Royal Veterinary and Agricultural University, Copenhagen, Denmark

The average soybean yield in Thailand has remained consistently low for many years, although high yielding local varieties have been adopted for cultivation. The reasons for this have been the disproportionate application of the fertilizer nutrients resulting in nutritional disorders and lack of a suitable method for diagnosing and controlling the nutritional disorders in the plants. In order to improve the situation, a new fertilization system based on the chemical composition of young plants, developed in Denmark, was examined for its applicability in Thailand.

The fertilization system consists of the following quantitative methods of correcting, evaluating and controlling the chemical composition and nutritional status of the plants: 1) correction of the chemical composition of the young plants to a fixed dry matter weight level—model niveau for all methods; 2) diagnosis of the nutritional status of the young plant based on its chemical composition at model niveau; 3) prognosis, based on diagnosis; 4) therapy based on diagnosis and prognosis with the aim of increasing the final yield and improving the chemical composition of mature plant parts; and 5) trophogenesis, i.e., determination of the nutritional status of the young plant based on production of final yield and chemical composition of mature plant parts. At this meeting, the results of the preliminary investigations on the development of diagnosis and prognosis models for soybean will be presented.

The models for dry matter correction and diagnosis and prognosis of soybean have been developed on the basis of field experiments conducted during the past three years with various levels of application of N and P. The models consist of curves for relationships between: 1) concentration of the nutrient and dry matter weight at various samplings, 2) concentration of the nutrient at a fixed dry matter weight level and final bean yield, and 3) optimal concentration and pure effect concentration of various nutrients. In the method of diagnosis and yield prognosis, the nutritional status for the various nutrients is expressed individually by absolute and relative deficiencies and excesses of nutrient concentrations and integrated by the criterion of the overall nutritional status, the final yield. The testing of the method has shown a high degree of reliability. This indicates a strong possibility for adoption of the system in Thailand for improving the production of soybeans.

INFLUENCE OF WATER STRESS ON THE PHOSPHORUS NUTRITION OF SOYBEANS

L. C. Campbell and D. Ridho, Department of Agronomy and Horticultural Science, University of Sydney, N.S.W. Australia 2006

Soybeans are being grown on a small scale under dryland conditions in N.S.W. These experiments were undertaken to obtain a preliminary assessment of the effect of water stress on the response of soybeans to applied phosphate. Both glasshouse and field trials are being done. The response to applied P and water stress in terms of several agronomic and chemical characteristics of soybeans (cv. Lee) was studied. The experimental design was a randomized complete block design of three replicates containing five levels of P (10, 35, 80, 200 and 300 kg P/ha), and two water stress (nonstressed and stressed) factorial combinations. The plants were grown in a pot experiment in a temperature controlled glasshouse where the day temperature was 24 C and the night temperature was 19 C. There were seven harvests during the pod filling period (0, 14, 28, 42, 56, 70 and 84 days after anthesis). Attributes studied included crop development, dry matter yield, P in the plant parts, and oil content in the seeds.

Applied P increased the dry weight of plant parts and grain yield. Water stress reduced this response to applied P. Phosphorus application increased both P content and concentration in all plant parts and seeds, but water stress reduced both of these parameters at all levels of P. The P content of the stem and pods was translocated into the seeds in the period between day 28 and day 84 after anthesis. Water stress tended to reduce the translocation of P from the stems and pods to the seeds at all levels of P. At the last harvest (day 84) the seed contained the highest concentration of P. Phosphorus application tended to increase the percentage oil in the seeds, but water stress did not affect this percentage. The major effect was thus only in terms of yield. There were very few interactions found between P level applied and water stress on the various parameters measured, i.e., growth components, P content, P concentration, oil content, or oil percentage.

MANGANESE NUTRITION OF SOYBEANS

D. P. Heenan, L. C. Campbell and O. G. Carter, Department of Agronomy and Horticultural Science, University of Sydney, N.S.W. Australia 2006

During initial screening of cultivars for growing under N.S.W. conditions, symptoms of Mn toxicity were observed. Furthermore, it was established that there was a differential sensitivity to Mn toxicity between cultivars (Carter et al., 1975). This work extends the initial findings. Four soybean varieties showed a range of tolerance to high Mn concentrations in solution culture. Seedling dry matter yields of Lee, Custer, Dare and Bragg were reduced by 27, 38, 42 and 59 percent, respectively, when the supply of Mn was raised from optimal (1.8 μM) to high (275 μM) concentrations. Dry matter yields of mature plants and final seed yields of Lee and Bragg grown in sand culture under a rain shelter were reduced by 275 μM maganese, the reductions being greater for Bragg than Lee. Manganese deficiency conditions reduced vegetative growth, seed yield and percentage oil content of seeds to a similar extent for both Lee and Bragg.

A study of a range of F_6 progeny of Amredo (a Mn tolerant variety) and Bragg parents and the F_2 progeny of Lee and Bragg parents suggested that tolerance to high Mn supply was controlled genetically. While data were not sufficient to make any realistic comment on the number of genes involved, the gene(s) imparting tolerance were at least partially dominant over the susceptible gene(s).

Reciprocal scion-stock grafts showed that the genotype of the shoot controlled varietal tolerance to high Mn concentration in solution. Transport and distribution studies indicated partial mobility or retranslocation of Mn in plants suffering from Mn deficiency. For plants grown in solutions containing 1.8 to 450 μM Mn levels high concentrations were accumulated in old leaves resulting in a decreasing concentration gradient between old and young leaves. The leaves contained higher concentrations than petioles which in turn were higher than the stems. However, there was no difference between varieties Lee and Bragg in the distribution of Mn to plant parts or in the concentration of Mn in actively growing tissue.

The rate of root absorption of Mn was reduced by increasing the solution concentration of either Ca, K, Fe or by decreasing the solution pH from 5 to 4. Retention of Mn in the roots was promoted by increasing solution levels of either Ca or Fe or by increasing the solution pH. Increasing the supply of Mn to 275 μM considerably reduced uptake of Mg but did not induce Mg deficiency symptoms in the present experiments. Iron deficiency symptoms were induced by 275 μM Mn when the Fe solution concentrations were reduced to low levels (2 μM). There was no effect of Mn on root absorption of Fe but high Mn levels reduced the transport of Fe from the roots to the tops.

PHOTOSYNTHESIS, RESPIRATION, CHLOROPHYLL AND LEAF RESISTANCE AS AFFECTED BY IRON DEFICIENCY

L. C. Campbell and P. D. Kemp, Department of Agronomy and Horticultural Science, University of Sydney, N.S.W. Australia 2006

In order to study a range of iron deficiencies the effect of both progressive iron deficiency and recovery from iron deficiency on soybean was measured. Soybeans (cv. Wayne) were grown in solution culture (pH 5.5) under glasshouse conditions (24/19 C). Four treatments were applied: nil Fe, 22.5 μM Fe sequestrene 138 (control), nil Fe until first trifoliate leaf developed marked deficiency symptoms and then 2.5 μM Fe sequestrene plus 50 μM $FeSO_4$, and a control for the penultimate treatment.

Harvests were made on the 14th, 16th, 19th and 21st day after sowing. On day 21 the iron concentration (μg Fe/g dry weight) of the control 1st trifoliates was 125, and that of deficient trifoliates was 38. The respective values for the 2nd trifoliate were 95 and 5. Gross photosynthetic rate, fresh weight basis, was lower ($p<0.05$) in the deficient leaves than in the control leaves. In contrast to this, gross photosynthetic rate, chlorophyll basis, was higher ($p<0.05$) in deficient leaves; except in the 2nd trifoliate on day 21, when it was lower ($p<0.05$). Chlorophyll concentration of deficient leaves ranged from 50 to 15 percent of control leaves. The 2nd trifoliate was more chlorotic than the 1st trifoliate. The respiration rate, fresh weight basis, was lower ($p<0.05$) in deficient leaves. The adaxial leaf resistance was greater in deficient leaves ($p<0.05$), but the abaxial leaf resistance was only significantly greater ($p<0.05$) on day 19. No effect on stomatal density was observed.

Iron was added 16 days after sowing. Harvests were made on the 15th, 17th, 18th, 19th and 20th day after sowing. Both the 1st and 2nd trifoliates had fully recovered from iron deficiency by day 20. The iron concentration, chlorophyll concentration, and gross photosynthetic rate, chlorophyll basis, did not fully recover until day 20. The gross photosynthetic rate, fresh weight basis, had recovered by day 17. The respiration rate had recovered by day 18, and leaf resistance by day 19. The chlorophyll concentration was most sensitive to iron deficiency. The leaves with a wide range of chlorophyll concentrations had normal gross photosynthetic rates, fresh weight basis, due to the increased efficiency of chlorophyll.

EFFECTS OF PHOSPHORUS ON SOYBEAN YIELD IN SOUTH WESTERN NIGERIA

N. O. Afolabi and O. A. Osiname, Institute of Agricultural Research and Training, Moor Plantation, Ibadan, Nigeria

The response of two soybean [*Glycine Max* (L.) Merrill] cultivars, Bossier and Improved Pelican, to applied phosphorus (P) was determined in two trials in the forest and Savannah zones of South Western Nigeria in 1077 and 1978. There was good response to P by both cultivars in forest soils developed from sedimentary rocks. Requirements being between 30 and 45 Kg P_2O_5/ha. In the Savannah zone, the response to applied P was small, the optimum being about 30 Kg P_2O_5/ha. The Bossier cultivar responded to phosphorus fertilizer better than improved Pelican. In general, the addition of phosphorus fertilizer to soybean plants has beneficial effects on the production of leaves, nodules, stems, total dry weights and seed yields. It was concluded that the addition of up to 30 Kg P_2O_5 to Savannah soils and 45 Kg P_2O_5 to forest soils of South Western Nigeria was adequate for good seed yield under the present cultural practices of the area.

ENGINEERING

Invited Papers

MECHANIZATION ALTERNATIVES FOR SMALL ACREAGES IN LESS DEVELOPED COUNTRIES

M. L. Esmay, Department of Agricultural Engineering, Michigan State University, East Lansing, Michigan 48824

Soybean production is labor intensive in the less developed countries of the world. Small fields that average less than one acre in many countries and the lack of investment capital for machinery have limited mechanization. Also, readily available labor has made mechanization for labor efficiency not only unnecessary but undesirable. Improved production and post production equipment and methods have been selectively introduced for the purpose of increasing yields and for minimizing crop losses. Seed and fertilizer placement machines, even though of the simple animal drawn type, have been found more reliable than hand placement with resulting increased yields. Improved sprayers for plant protection from insects and diseases have also proved justified on the basis of increased yields. Where and when labor is available, harvesting can be done quite adequately by hand. Threshing by hand beating can, however, result in excessive broken beans if the moisture conditions are right. Field shattering can be a problem with hand sickle cutting and field drying, although in some countries the individual beans are picked up one by one and salvaged. Soybeans are grown during the cooler season in some sub-tropical countries following one or two crops of rice during the hot season. The International Crop Research Institute for the Semi-Arid Tropics (ICEISAT) in India is developing a farming system for dry land crops. It utilizes mainly animal power and a permanent ridge and ditch system. The narrow ditches are 1½ meters on center and serve as paths for the animals and implement wheels, in order to prevent soil compaction in the plant growing ridge area. The ditches may also be used for irrigation.

SOIL CONSERVATION PRACTICES IN SOYBEAN PRODUCTION

J. C. Siemens, University of Illinois, Urbana, Illinois 61801

Soil erosion is a serious problem with soybean production especially on sloping land. Conservation tillage is being promoted as a means of effectively reducing soil erosion. It is a concept based on efficient row-crop production with adequate control of soil erosion caused by wind and water. The primary concern is for soil conservation, but water and energy conservation may be added benefits. A rainfall simulator was used in Illinois to compare erosion characteristics of different tillage systems after growing soybeans. The first 4 inches of water applied caused a soil loss of 8 tons per acre with fall plowing, 2.5 tons per acre with chisel plowing, and 1 ton with no-tillage. Even with the significant reduction in soil erosion with conservation tillage, farmers are not using such systems on a large scale. The reasons are because yields are not increased and costs are not decreased significantly. Evidently, tillage improves soil characteristics for crop growth and helps to control pests (weeds, insects, and diseases). Additional information is badly needed on the interaction of tillage and chemicals on pest control. The effectiveness of conservation tillage in erosion control is so great that we must strive to perfect the concept so that it becomes an acceptable technique for most farmers.

EQUIPMENT FOR NARROW-ROW AND SOLID-PLANT SOYBEANS—"ONE STEP FORWARD"

R. I. Throckmorton, International Harvester Company, Chicago, Illinois

New equipment for solid plant to narrow row soybean culture, commercial equipment and farmer designed equipment to meet culture needs, and the adequacy of current descriptions of systems for farmer interpretation, solid plant versus very narrow row, will be discussed.

SOYBEAN SEEDCOAT DAMAGE DETECTION

M. R. Paulsen, Agricultural Engineering Department, University of Illinois, Urbana, Illinois 61801

Soybean seedcoat damage is an important indicator of present soybean quality and future quality retention. High levels of seedcoat cracks have been associated with increased fungal growth leading to a reduced storage life, reductions in oil yield and quality, and increased percentages of splits during handling. Three tests for detecting seedcoat crack percentage were conducted on soybeans containing different levels of seedcoat cracks. These tests included an improved indoxyl acetate test, a tetrazolium test, and a sodium hypochlorite test. The indoxyl acetate test detected higher levels of seedcoat cracks than the other two tests. Soybeans found to have no seedcoat cracks also were found to have higher warm germination percentages than those beans with seedcoat cracks. Warm germination percentages of soybeans with no seedcoat cracks were not affected adversely by the indoxyl acetate test. In seedlots with very low levels of damage, statistically significant differences between seedlots were often difficult to show because of relatively small sample sizes and sampling variability. In such cases comparisons were made between indoxyl acetate tests, stein breakage tests, and mechanical sieving analyses which provided other important indications of soybean quality.

DRYING METHODS AND THE EFFECT ON SOYBEAN QUALITY

G. M. White, I. J. Ross, and D. M. Egli, University of Kentucky, Lexington, Kentucky 40506

The various types of grain dryers and their application to soybean drying are discussed. A review of literature related to soybean drying is presented along with a discussion of how heated air drying affects soybean quality. The relative merits of each drying method are discussed along with operational guidelines for reducing potential drying damage. Research by the authors on the levels of seedcoat and cotyledon damage in soybeans dried with heated air is described. In thin-layer drying experiments the development of both types of damage was found to be highly related to the relative humidity of the drying air and the initial moisture content of the soybeans. Drying air temperature did have an affect on damage but it was not as significant. Seedcoat cracks did not approach zero until the relative humidity of the drying air increased to approximately 50 percent. Little or no increase in cotyledon cracks occurred in the soybeans when the relative humidity of the drying air was above: 19 percent for 16 percent initial moisture; 25 percent for 20 percent initial moisture; and 35 percent for 24 percent initial moisture.

In a laboratory batch-type dryer soybeans of 19 and 25 percent have been dried with air flow rates of 0.1 and 0.2 $m^3/s{\cdot}m^2$ and drying air temperatures of 50 and 65 C. Results indicate that drying damage (seedcoat and cotyledon cracks) varies significantly with position in the dryer. Maximum damage at different locations was found to differ significantly even after the drying front had passed completely through the depth being dried. Damage was lower the further the beans were from the air inlet. Drying air temperature and air flow rate were found to significantly affect the magnitude of the damage gradient in the bin at the completion of drying. The significance of the above results relative to the practical operation of soybean dryers is discussed. Research needs are summarized.

SOYBEAN SEED STORAGE UNDER CONTROLLED AND AMBIENT CONDITIONS IN TROPICAL ENVIRONMENT

E. J. Ravalo, Agricultural Engineering Department, University of Puerto Rico, Mayaguez, Puerto Rico 00708 and E. D. Rodda, F. D. Tenne, and J. D. Sinclair, Agricultural Engineering Department, University of Illinois, Urbana, Illinois 61801

Soybean seed produced in Puerto Rica was stored in four types of containers (sealed cans, plastic bags in unsealed cans, fertilizer bags and cloth bags) at three moisture contents (3, 10, 13 percent dry basis) for three storage periods (3, 6, 9 months). Parallel tests in sealed metal cans at Puerto Rican mean ambient temperature were conducted in Urbana, Illinois. Results of storage studies demonstrated that maintaining low initial moisture content was essential for successful storage. Storage in sealed metal cans at a constant mean ambient temperature gave results similar to those for ambient temperature storage.

Contributed Papers

EFFECT OF ROW WIDTH, TYPE MACHINE, TIMELINESS AND SPEED ON HARVEST LOSSES OF DETERMINANT SOYBEANS

M. M. Mayeux and J. Gregg Marshall, Louisiana State University, Baton Rouge, Louisiana

The row crop header and the flexible floating cutter bar header were used to harvest Forrest, Davis and Bragg soybeans. The harvest was conducted at maturity and at one week intervals for the following three weeks. Spacings of 18 cm, 76 cm and 102 cm were used. The effect of speeds of 2.9, 4.6, 6.2 and 7.3 km per hr on losses were evaluated with both headers at a row spacing of 102 cm. Wider rows showed significantly higher total harvest losses. Losses increased as speeds above 4.6 km per hour were used. There was no significant difference in the losses with the row crop header as compared to the flexible floating cutter bar. Benlate treatment of soybeans for disease control delayed maturity four days. There was no significant difference in the harvest losses experienced when harvesting treated soybeans. When all varieties and all harvest dates were considered the mean loss when harvesting with the row crop header was 4.44 percent whereas the loss with the floating flexible cutter bar was 4.41 percent. The row crop header requires a fixed row width, wider headlands and more careful driving. It is more difficult to operate under muddy field conditions due to the necessity for precision driving.

NEW COMBINE THRESHING DESIGNS AND SOYBEAN SEED QUALITY

G. H. Diener and D. M. Byg, Agricultural Engineering Department, Ohio State University, Wooster, Ohio

A new design combine (twin rotor-threshing grate) and a conventional design combine (cylinder-concave) were tested to compare the effect of their threshing mechanisms on soybean seed quality. Tests were performed on three soybean varieties at two seed moisture levels and three combine threshing speeds. Samples of soybean were collected during each test run and analyzed to determine percent germination, percent splits, percent cracks, breakage (Stein Breakage Test), and percent vigor (Tetrazolium Test). Preliminary results indicate that for most test runs, soybeans harvested with the new design combine had better germination, fewer splits, less cracks, less breakage and higher vigor. Soybean loss from the rear was less for the new design combine than for the conventional design combine.

NITROGEN FIXATION

Invited Papers

NITROGEN FIXATION IN SOYBEANS

R. W. F. Hardy, U. D. Havelka, and P. G. Heytler, Central Research and Development Department, E. I. Du Pont de Nemours and Company, Wilmington, Delaware 19898

The research frontier on soybean N_2 fixation will be synthesized proceeding from the molecular level (nitrogenase, Fe-Mo cofactor mechanism, ancillary proteins, enzymes, products of ammonia incorporation, sources of energy, location, mapping, regulation, and transfer of *nif* and related genes) to the cellular level (asymbiotic N_2 fixation, specificity, lectins and infection) to the legume microsymbiont association (measurement technique including "A" value, biological energy cost, CO_2-enrichment, H_2 production and re-utilization, senescence, fixed nitrogen, and foliar fertilization). Opportunities for improvement based on biochemical, genetic, physiological, and agronomic limitations of the soybean symbiosis will be identified. The cost of N_2 fixation by comparison of root respiration of nodulated and non-nodulated soybean isolines is about 20 kg carbohydrate per kg N_2 fixed and this value will be compared with those of 5-20 for soybean and other legumes determined by other methods. These values are 1.25 to 5.0 times the theoretical value for the highly inefficient nitrogenase enzyme. Increasing biological energy available for N_2 fixation or increasing the efficiency of N_2 fixation is a mjaor objective in soybean N_2 fixation research.

RHIZOGENETICS OF SOYBEAN

R. W. Zobel, USDA/SEA, Cornell University, Ithaca, New York 14853

Rhizogenetics is the study of genetic control over plant below ground structures. There is little information in reference of rhizogenetics of soybean. With the possible exception of the few non-nodulating genes which have been discovered, and the occasional reference to differences in nutrient uptake, gross root morphological differences, or fluorescence of roots, this area has been little studied. Nitrogen fixation is a symbiotic relationship where the complex genomes of two extremely different organisms 'cooperate' to produce the final product, a nodule capable of fixing atmospheric nitrogen. The word 'cooperate' is critical to a clear understanding of the process recognized as nitrogen fixation. There has been and currently is extensive research on the genetics of the rhizobium symbiont, including genetic engineering via recombinant DNA techniques in attempts to improve the bacterium.

Research on the plant host is certainly not as well developed, nor is it even fractionally as well supported. Techniques in use currently involve destructive sampling to measure rates of fixation of differing genetic lines, to count and measure nodules, and to study their anatomy and its relation to normal processes in the plant. These approaches are yielding information which indicates the presence of a wealth of variability in soybean germplasm collections. One inescapable conclusion is that the plant may exert as much as or more genetic control over the whole system as the rhizobium does. Plant root structure plays a strong role in the patterns of nodulation, but there appears to be an over-riding control of nodulation patterns by the plant in addition to those conditioning rooting patterns. Techniques are being developed to measure rates of nitrogen fixation under field conditions nondestructively. Preliminary data from these studies indicate that only through continuous sampling of the same plant can 'true rates' of nitrogen fixation be monitored and determined.

Results from these studies should lead to information which can be applied to the improvement of the plants share of the genetic information directing nodulation and nitrogen fixation. An additional approach is critically necessary. This is the study of how specific strains of rhizobia balance with specific plant strains. There are many studies approaching specific aspects, such as the host recognition sites. Breeding both host and symbiont, in concert, holds the greatest chance for success, especially if we are to utilize the advances possible with genetic engineering of the symbiont.

Only through extensive rhizogenetic studies can we hope to bring our knowledge of the host genetics up to the level necessary to provide a basis for truly improving nitrogen fixation and possibly reducing the worlds demand for artifically fixed nitrogen.

RHIZOBIA AND SOYBEAN PRODUCTION

J. C. Burton, Research and Development, The Nitragin Co., Inc., Milwaukee, Wisconsin

Soybean culture continues to expand into alien countries, virgin soils and unaccustomed climates. Success in the program depends not only on prudence in selecting the host genotype, but also in the selection and use of the microsymbiont, *Rhizobium japonicum*, which produces the nodules and supplies the nitrogen for seed development.

Strains of *R. japonicum* must be selected not only for their N_2 fixing ability on specific host genotypes, but for their ability to survive both on the seed and in the soil under the stress of low moisture and high temperatures in sufficient numbers to colonize the root, effectively nodulate the young seedling, and work efficiently to provide their host with adequate nitrogen. In many cases, the rhizobia may have to work effectively even in phosphorus-deficient soils containing moderately high concentrations of soluble aluminum and manganese.

Factors such as day-length, air and soil temperature, soil reaction and nutrient level may be simulated sufficiently well in the growth chamber or phytotron to distinguish the poor prospects among strains of *R. japonicum.* However, the ultimate proof of true effectiveness in the microsymbiont can be obtained only under the stress of fruiting plants growing under natural soil and climatic conditions competing with the native weeds, insects and diseases.

Another important consideration is the delivery system for the nodule bacteria. Highly effective strains of *R. japonicum* can be of little benefit unless they can be established in the rhizosphere of the developing young seedling in sufficient numbers to assure that they will produce a majority of the nodules. Inoculation methods considered adequate in temperature climates, with normally adequate moisture, moderately fertile soils, and a favorable pH would certainly not be acceptable under the severe stresses prevalent in many of the tropical and subtropical countries. Delivery systems employing large inocula and application separately from the seed will be described and evaluated.

INFLUENCE OF SOYBEAN SEEDLING VIGOR ON NODULATION

R. S. Smith and M. A. Ellis, Department of Agronomy, University of Illinois, Urbana, Illinois 61801 and Department of Crop Protection, University of Puerto Rico at Mayaguez, Mayaguez, Puerto Rico 00708

Nodulation establishment in soybeans [*Glycine max* (L.) Merr.] is influenced by environmental, rhizobial and host symbiont factors. The objective of this study was to determine the effect of individual soybean seedling vigor on the establishment of tap root nodules, total nodules and nodule dry weight per plant. The vigor of three soybean cultivars was decreased using the accelerated aging technique by placing seeds in an incubator at 40 C and 100 percent relative humidity. Subsamples for each cultivar were removed from the aging chamber after 1, 2, 3 and 4 days and blended with non-aged seed to provide seed lots with a high degree of variation in seedling vigor.

Sandbench and field experiments were conducted with seeds planted at a uniform depth and treatments established by tagging individual seedlings according to the number of days required for each seed to emerge. Uniform inoculation was accomplished by applying a single strain liquid culture of *Rhizobium japonicum* onto the seeds in the open furrow. The weights of plant tops, roots and nodules, and numbers of tap roots and total nodules were evaluated at 26 days in the sandbench, 40 days in the first field experiment, and 27 and 38 days in a second field experiment. Significant decreases in the numbers of tap root nodules and total numbers of nodules were observed between each of the three earliest emerging treatments in a sandbench trial. In both sandbench and field experiments, there were significantly more tap root nodules, total numbers of nodules and nodule dry weights per plant on the earliest emerging treatment than on the latest emerging treatment. Decreases in nodulation parameters were observed with increasing days to emerge. All trials had a significant (.01 level) negative correlation between days to emerge and plant top weights and root weights.

Results of this study suggest that a reduction in seedling vigor has a significant effect on quantity and distribution pattern of soybean nodulation. Therefore, plants of uniform vigor should be utilized in experiments or tests evaluating soybean nodulation. This may assist in reducing undesirable variables in complex studies involving the symbiotic relationship between soybean and *Rhizobium japonicum.*

Contributed Papers

SUCROSE METABOLISM IN DETACHED SOYBEAN NODULES

I. K. Stovall, Department of Agronomy, University of Illinois, Urbana, Illinois 61801

Sucrose is the principal carbon compound transported from the shoot to the root of the soybean plant. The work described below was conducted to determine the fate of sucrose carbon in intact nodules. Whole soybean nodules with a small piece of root attached were removed from the roots and surface-sterilized with NaClO. The ^{14}C-substrate was added, the nodules were incubated and respiratory CO_2 was collected continuously. At selected intervals, the nodules were fractionated to determine the fate of ^{14}C from the added substrate. Most of the radioactivity from ^{14}C-sucrose was found in the cytosol of plant cells in the nodule interior, with very little radioactivity in nodule cortex cells or in the bacteroids. Only a limited amount of ^{14}C-CO_2 was evolved by whole nodules. Metabolism of intermediates of sucrose metabolism (e.g., succinate) was much slower with whole nodules than with isolated bacteroids obtained from an identical weight of whole nodules. The results suggest that bacteroids *in situ* are not able to compete effectively with the plant cells of the nodules for energy-rich metabolites, and, therefore, the inability of the bacteroids to obtain energy sources within the nodule may limit nitrogen fixation.

EFFECT OF LIGHT ENHANCEMENT ON NITROGEN ASSIMILATION AND DRY MATTER PRODUCTION BY FIELD GROWN SOYBEANS

L. E. Schweitzer and J. E. Harper, USDA and Department of Agronomy, University of Illinois, Urbana, Illinois 61801

In field environments, soybeans develop a closed canopy which may reduce the penetration of light to leaves at the lower nodes. Canopy closure becomes especially significant during the later (pod-fill) stages of plant growth. The objectives of the current study were to measure the effects of lower canopy light enhancement on: 1) root nodule activity (acetylene reduction), 2) leaf nitrate reductase activity, and 3) dry matter accumulation and seed production by field grown soybeans (*Glycine max* L. Merr. cv. Williams).

Light enhancement in treated plots was achieved through the placement of foiled reflectors at 45° angles on either side of the treated rows. Appropriate check plots (blackboards and wire mesh screens placed at 45° angles on either side of the treated rows) were included to separate temperature and light effects in the light enhanced environment. An additional treatment consisted of accelerated floral induction. This treatment was designed to move the reproductive growth period forward in the season to receive greater solar radiation.

In the field environment, light enhancement by reflectors produced a marked increased in nodule activity (acetylene reduction), and also delayed nodule senescence. In addition, nitrate reductase activity (*in vivo*) was increased in the leaves of light enhanced plants relative to controls. Total dry matter accumulation, seed production, and the grain to stover ratio were also increased for soybeans grown in the light enhanced environment.

Early flowered plants produced a profile of nodule activity (acetylene reduction) which was compressed in the season, reflective of early senescence relative to nodules of control plants. As with control plants, peak nodule activity (acetylene reduction) was observed during the respective podfill periods. Peak nodule activity (acetylene reduction) for the early flowered plants was well above that measured for control plants of equivalent chronological age and comparable nodule mass. Therefore, peak nodule activity in these early flowered plants was associated with a marked increase in specific activity, relative to controls.

Results of this investigation suggest that limited light penetration in a closed canopy environment may limit both N_2-fixation (acetylene reduction) by root nodules and nitrate reduction (*in vivo* nitrate reductase) by leaves of field grown soybeans. Closed canopy shading may also reduce dry matter accumulation and seed production. Peak nodule activity measured for early flowered plants was associated with a rise in specific activity during podfill. This rise in activity appears to occur independent of nodule mass or chronological age.

EFFECTS OF INOCULATION ON SOYBEAN NODULATION, N_2 FIXATION AND SEED YIELD

H. L. Peterson, Department of Agronomy, Mississippi State University, Mississippi State, Mississippi

Soybean inoculation is an accepted management practice throughout the world, especially during initial crop introduction. Few studies have reported increased soybean seed yield through inoculation in soils containing established populations of soybean rhizobia. This study was undertaken to determine if inoculation at extremely high rates could promote nodulation by superior, N_2-fixing strains of *Rhizobium japonicum*, thus increasing N_2 fixation and soybean seed yield.

Lee soybeans were planted in Leeper soils at several Mississippi locations in 1976-78. Seeds were inoculated at planting by dribbling aqueous suspensions of rhizobia in the seed furrows. Plant growth was monitored throughout each growing season. Plants were sampled at the R1-R2 stage of development. Samples were separated into roots, nodules, stems, leaves + petioles and pods. Nodules were weighed and twenty nodules/plot were analyzed serologically to determine the extent of nodulation by the inoculated strains. Total N in various plant parts was determined by Kjeldahl digestion to include nitrate.

Inoculation with as many as one billion rhizobia/seed (3 cm of row) resulted in only 10 to 20 percent nodulation by the inoculated strain. This strain, however, increased the plant nitrogen content 42 kg/ha. Successful inoculation with a superior strain of rhizobia increased seed yield an average of 2.0 q/ha. Inoculation at extremely high rates can successfully promote nodulation by superior, N_2-fixing strains of *R. japonicum.* New inocula or inoculation procedures must be developed to increase nodulation by superior strains of rhizobia in soils under adverse conditions of moisture and temperature.

ENHANCEMENT OF NODULATION OF *GLYCINE MAX* BY MIXED CULTURES OF *RHIZOBIUM JAPONICUM* AND *AZOTOBACTER VINELANDII*

T. A. Burns, P. E. Bishop and D. W. Israel, Departments of Microbiology and Soil Science, North Carolina State University, Raleigh, North Carolina 27650

Under greenhouse conditions it was observed that inoculation of *Glycine max* var. Ransom with mixed cultures of *Rhizobium japonicum* strain 61A76 and *Azotobacter vinelandii* strain OP gave increased numbers of nodules as compared to inoculation with *R. japonicum* alone. A survey of the literature revealed that this phenomenon had previously been observed, but that it was poorly characterized. Hypotheses which could explain the enhanced nodulation are: production of stimulatory-substances by *A. vinelandii*; degradation of substances inhibitory to nodulation; and/or provision of low levels of combined nitrogen through nitrogen fixation by *A. vinelandii.* To test the latter possibility, plants were inoculated with mixed cultures of *R. japonicum* and mutant strains of *A. vinelandii* unable to fix nitrogen (Nif^-). After growth under nitrogen-free conditions, nodulation enhancement was observed in the presence of both Nif^- and Nif^+ *A. vinelandii* strains. Thus, nitrogen fixation cannot be the primary cause of nodule enhancement. When heat-killed azotobacter cells were applied in mixed inocula, nodulation was not increased above the treatment with *R. japonicum* alone. This result suggests that viable azotobacter cells are essential for increased nodulation. Enhanced nodulation was also observed with the *Trifolium repens-Rhizobium trifolii* and the *Vigna unguiculate-Rhizobium sps.* symbiotic associations, indicating that this phenomenon is not limited to the *G. max-R. japonicum* symbiosis.

STUDIES ON MOLYBDENUM APPLICATION TO SOYBEAN

C. S. Weeraratna, Faculty of Agriculture, University of Sri Lanka, Peradeniya, Sri Lanka

Molybdenum is known to be essential for the activity of nitrogenase, an enzyme which is involved in nitrogen fixation by root nodule bacteria. Availability of molybdenum in soil tends to increase with pH. It could be applied to soybean in the form of sodium or ammonium molybdate as a seed treatment or as a soil application. Investigations were carried out to examine the most effective method of applying molybdenum to soybean grown in a Reddish Brown Lateritic soil of pH 5.3. In a preliminary experiment conducted in pots, molybdenum as ammonium molybdate (2 percent solution) was applied: 1) to seeds inoculated with nitrogen culture, 2) to soil as a spray (5 oz per acre) at sowing, 3) one week, or 4) two weeks after sowing inoculated seeds. Dry weight and nitrogen content of plants and yield data indicate that seed treatment was the most effective.

A field experiment was conducted in a similar soil to examine the best concentration of ammonium molybdate to be used as a seed treatment. Three concentrations, viz. 1, 2, and 5 percent, were tested. Yield data and nitrogen contents of plants show that the 2 percent solution was significantly better than 1 percent. The 5 percent solution did not show a significant difference from the 2 percent solution.

In another experiment conducted to confirm the exact role of ammonium molybdate, a 2 percent solution of this salt was applied to inoculated and non-inoculated seeds. Yields and nitrogen contents of inoculated plants were significantly higher than the non-inoculated plants and the control. There was no significant difference between the control and non-inoculated plants treated with ammonium molybdate, showing that the beneficial effects of applying ammonium molybdate was due to its influence on nitrogen fixation.

SELECTION OF SOYBEAN CULTIVARS CAPABLE OF FORMING AN EFFECTIVE SYMBIOSIS WITH RHIZOBIA INDIGENOUS TO AFRICAN SOILS

E. L. Pulver, International Institute of Tropical Agriculture, PMB 5320, Ibadan, Nigeria

Numerous experiments conducted in the tropics have demonstrated that high yielding soybean cultivars (primarily of U.S. origin) require inoculation with *R. japonicum* to achieve their full yield potential. However, many other grain legumes nodulate with indigenous rhizobia and responses to inoculation are seldom observed. This work was done to identify soybeans capable of forming an efficient symbiosis with native rhizobia thereby circumventing the need for an inoculant. All the experiments were conducted on soils not previously cultivated to soybeans and deficient in nitrogen.

In 1977, field trials were established in two locations in Nigeria using six cultivars grown with and without an inoculum of *R. japonicum.* Three of the cultivars (all of S.E. Asian origin) nodulated profusely with indigenous strains whereas the other three (improved material from the U.S.) nodulated poorly. The yield of the three S.E. Asian cultivars was unaffected by inoculation whereas the U.S. material showed a 61 percent yield increase. Similar results were obtained in greenhouse pots using uncontaminated soil from a farmer's field.

Shoot-root grafts demonstrated that the symbiosis with indigenous rhizobia is efficient. The shoot of a high yielding U.S. cultivar (Bossier) grafted onto the root of a S.E. Asian cultivar in uninoculated field soil produced yields equal to the U.S. material grafted onto itself and inoculated.

Screening for soybean cultivars able to nodulate with a wide range of indigenous rhizobia was initiated in 1978. Two hundred and fifty cultivars of diverse origin were planted in six locations in Nigeria. Testing sites ranged from the high rainfall, acid soil regions to the semi-arid savanna. Only eight cultivars nodulated profusely with effective indigenous rhizobia at all sites. Present work centers on incorporating this promiscuity into agronomically superior cultivars.

RHIZOBIUM STRAIN INTERACTIONS WITH HEALTHY AND INFECTED SOYBEAN CULTIVARS

T. M. Carmody, P. J. McGinnity and G. Kapusta, Plant and Soil Science, Southern Illinois University, Carbondale, Illinois

Studies demonstrating characteristics such as ineffective nodulation and inefficient N_2-fixation have shown the need for continued research in the area of *Rhizobium japonicum* strain x soybean cultivar interaction. A study was designed to evaluate strain competiveness and N_2-fixing efficiency of three commercial soybean cultivars, and their parents, when inoculated with several *R. japonicum* strains commonly found in soybean fields.

Twenty-two soybean cultivars were grown in modified Leonard jars of sterile sand, in growth chambers. $N_2(C_2H_4)$-fixation was determined at the end of five weeks. Significant differences were found between cultivars and strains 110 and 123, but not 138. Strain 110 gave the highest total N_2-fixation and specific N_2-fixation with all cultivars when compared to strains 123 and 138. Values associated with strain 123 plants were somewhat higher than those of 138, but still considerably lower than those of 110. A general trend of high fixation values occurred with cultivars not currently grown in the field, particularly when associated with strain 110. Strain 110 was shown to form 80 percent or more of the nodules, with each cultivar when mixed with strains 123, 138 and others.

The spread of the soybean cyst nematode (SCN) has necessitated the breeding of more SCN resistant cultivars. The SCN is thought to decrease soybean yields by reducing nodulation and N_2-fixation. A study was designed to determine interaction effects of soybean cultivars (SCN resistant and susceptible), *R. japonicum* strain, and SCN races on nodulation and N_2-fixation.

Four soybean cultivars, three SCN resistant and one susceptible, were grown in sterile sand culture and inoculated with possible combinations of SCN races 3 and 4 and *R. japonicum* strains 3, 110 and 123. Nodule mass and $N_2(C_2H_4)$-fixation were determined after six weeks. A significant cultivar x strain interaction was observed in total and specific N_2-fixation and nodule mass. Strain 110 inoculated Forrest and Bedford were superior to strain 110, 123 or 3 inoculated Franklin and Williams in N_2-fixation and nodule mass. A significant cultivar x SCN interaction indicated that race 3 was antagonistic to all cultivars but Bedford in N_2-fixation. Race 3 reduced nodule mass on all cultivars while race 4 increased nodule mass and caused no reduction in N_2-fixation.

The variability in cultivars x strain N_2-fixation and factors which affect it should be exploited by soybean breeders to maintain maximum nodulation effectiveness and N_2-fixing efficiency under variable conditions.

CONTRIBUTION OF SYMBIOTICALLY FIXED NITROGEN BY SOYBEAN AND OTHER GRAIN LEGUMES TO THE ASSOCIATED AND SUCCEEDING CROP

J. N. Singh and S. P. Kashyap, G. B. Pant University of Agriculture and Technology, Pantnagar, India

Studies were initiated to find out the contribution of symbiotically fixed nitrogen by soybean, black gram, green gram and cowpea to the associated crop and residual effect of soybean on succeeding crops of wheat and corn. The nodulating (inoculated) and non-nodulating isolines of Lee (soybean) green gram, black gram and cowpea were planted in mixed as well as pure stands. The non-nodulating soybean was flanked on both sides by two and three rows of nodulating soybean and inoculated grain legumes. In addition, non-nodulating soybean was grown at varying levels of N (20, 40, 80, 160 and 320 kg N/ha). To find out the residual effect of soybean on succeeding crops several double cropping sequences involving cereal crops and grain legumes were studied from 1972 to 1976. The yield of non-nodulating soybean in association with nodulating soybean and grain legumes increased 23 to 27 percent over non-nodulating pure stands. Non-nodulating soybean derived the greatest benefit in association with black gram (47 percent increase) followed by nodulating soybean (40 percent increase). The contribution of nodulating soybean to the associated crop was equivalent to the application of 80 kg N/ha to non-nodulating soybean. The protein content and the other yield contributing attributes of non-nodulating soybean were also benefited with the association of nodulating soybean and grain legumes. The response of non-nodulating soybean to N application was found up to 320 kg/N ha and the grain yield was *at par* with that of inoculated nodulating soybean (without applied nitrogen). The wheat yield following inoculated soybean was 34 to 47 percent higher than the wheat yield obtained following inoculated soybean. Among the different double cropping sequences, the highest yield of wheat was obtained from a soybean-wheat cropping sequence. The differences in corn yields following several grain legumes and soybean were not significant implying that soybean was not inferior to other grain legumes in respect of nitrogen economy of the succeeding crop. The soybean crop added about 20-30 kg additional available nitrogen to the succeeding crop.

ENTOMOLOGY

Invited Papers

SYSTEMS APPROACH TO PEST MANAGEMENT IN SOYBEANS

J. L. Stimac, Department of Entomology and Nematology, University of Florida, Gainesville, Florida 32611

A systems approach to pest management in soybeans places an emphasis on understanding all pests within the soybean agroecosystem. Dynamics of insect pests must be understood at various levels of spatial and temporal resolution. Management strategies based solely on within field changes in pest populations are strongly oriented toward treating symptoms of problems, not elucidating causes. Techniques of systems analysis and computer simulation can provide tools needed for more hollistic approaches to pest management in soybeans. Model construction and analysis should complement not circumvent experimental studies. Limitations of using simulation models for evaluating pest management strategies are rooted in poor quantitative descriptions of numerous control tactics (chemical, biological and cultural). Production goals heavily influence viability of pest management strategies. More emphasis should be placed on stability of production and less on optimum yield in any given year. Changes in economic systems may be necessary if we are to achieve long-term stable production of soybeans.

INTERACTION OF CONTROL TACTICS IN SOYBEANS

L. D. Newsom, Department of Entomology, Center for Agricultural Sciences and Rural Development, Louisiana State University, Baton Rouge, Louisiana 70803

Rapid changes in agriculture, especially accelerating dependence upon use of chemical pesticides for crop protection, have created environments that favor interactions between control tactics. Several potentially serious problems resulting from such interactions have been identified in soybean ecosystems. None has received sufficient attention from researchers. All are poorly, or not at all, understood. The following are some of the more important.

Prophylactic treatment of soybeans with chemicals having both insecticidal and nematicidal activity is practiced widely. They are often devastating in their effects on predators and parasites that usually regulate populations of pest insects below economic injury levels. Unauthorized use of toxaphene for control of sicklepod affects predators and parasites adversely. Also, it contaminates soils and bodies of water with residues that are accumulated to high levels in fish and other non-target species. Where cotton and soybeans are intercropped and organochlorine insecticides, especially endrin, are used for control of cotton pests, illegally high residues are found frequently in soybeans.

The fungicide benomyl is applied to huge acreages of soybeans for control of fungal pathogens. It adversely affects populations of entomopathogenic fungi that are important natural control agents of lepidopterous pests. Some insecticides are known to have fungicidal properties but the significance of their possible effects on entomopathogens is not known. Some herbicides, the triazines for example, used for weed control in corn may persist in soils at sufficient levels to damage, or destroy, stands of soybeans planted in the rotation. The significance of these interactions to the soybean industry will be discussed.

PEST PROBLEMS OF SOYBEAN AND CONTROL IN NIGERIA

M. I. Ezueh and S. O. Dina, National Cereals Research Institute, Ibadan, Nigeria

(See Addendum.)

STATUS OF INSECT PEST MANAGEMENT IN INDONESIA

D. Soekarna, Central Research Institute for Agriculture, Bogor, Indonesia

Soybeans in Indonesia occupy the fifth place among the foodcrops. From 1973 to 1976 the average annual harvested acreage was 724,902 hectares with a total production of 550,523 tons and an average yield of 7.6 quintals per hectar. Low production was attributed primarily to insect damage. More than 20 species of arthropod pests were recorded, among which the most important were: the beanfly, *Agromyza phaseoli* Coq., the podsuckers, *Riptortus linearis* F., and *Nezara viridula* L., and the podborer, *Etiella zinckenella* FN. Insects identified as important were: the leaffeeder, *Prodenia litura* F., *Plusia chalcites* Esp., *Lamprosema indicata* F., and the leaffolder, *Stomopteryx subsecivella* Zell.

Insect control is being done primarily using insecticides. Experiments using insecticides gave good control of the pests and gave economic yield responses, if applied properly and correctly. Tryazophos, pyridaphenthion, cyanophenphos and a combination of fenitrothion and cyanophenphos applied at 10, 30, 50 and 70 days after planting yielded respectively 1258 kg, 1179 kg, 1276 kg and 1262 kg per hectare, while in the untreated check the yield was 114 kg per hectare. The application at 30 and repeated at 50 and 70 days after planting gave a very low yield with the same insecticides. The very low yield was due to the very high beanfly infestation at the early seedling stage, which need early insecticide applications.

Efforts to employ pest management schemes instead of relying completely on insecticides was initiated in 1973. Experiments included a study on the seasonal occurrence of the major insect pests in relation to the growth stages of the plant in the soybean centers in Central Java and East Java. Screening to obtain resistant varieties or resistant gene sources for breeding material against the beanfly, the leafbettle and the podsucker was implemented. More than 200 varieties were screened for the beanfly, 100 varieties for the leafbettle and 215 varieties for the podsucker, *R. linearis,* but no single variety showing resistance to the insects was obtained.

A study on the susceptible and damage intensity of various seedling stages to the beanfly revealed that 5 to 10 days after planting was the most suitable period for egglaying and at this period heavy damage was recorded, ranging from 50 percent at 5, 56 percent at 6, 50 percent at 7, 46 percent at 8, 24 percent at 9, and 26 percent at 10 days after planting. Based on this data the insecticides should be applied at 5 to 10 days after planting to avoid serious damage. An experiment to develop an economic damage threshold to the podsucker was conducted using population densities ranging from 0, 2, 4, 6, and 8 insects per 9 hills and the damaged of pods of 0, 31.7, 46.2, 63.4, 83.8 and 92.3 percent, respectively. Studies on the economic damage threshold of the leafbeetle are in progress. Strategy of ongoing and future research on soybeans is directed toward the establishment and implementation of the integrated pest management system.

SOYBEAN INSECT PROBLEMS IN INDIA

A. K. Bhattacharya, G. B. Pant University, Pantnagar, India

Abstract not available at press time.

INSECT PROBLEMS ON SOYBEAN IN THE UNITED STATES

M. Kogan, Office of Agricultural Entomology, University of Illinois and Illinois Natural History Survey, Urbana, Illinois 61801

Throughout the nearly 23.2 million hectares planted to soybean in the U.S., an estimated 95 percent of the total amount of insect damage is produced by no more than eight species. Seven of these species belong to three major guilds: 1) lepidopterous defoliators–*Anticarsia gemmatalis, Pseudoplusia includens,* and *Plathypena scabra;* 2) coleopterous defoliators–*Epilachna varivestis* and *Cerotoma trifurcata;* and 3) pod feeding Pentatomidae–*Nezara viridula* and *Acrosternum hilare.* The eighth species–*Heliothis zea*–is locally important as either a foliage or a pod feeder. Among the various soybean growing regions in the U.S. there are vast differences in the frequency and amplitude of yearly population fluctuations. In addition some 20 individual species and complexes of species may be locally damaging from time to time. Extensive surveys conducted in the last 30 years in the U.S. show that soybean insect communities have not reached yet a state of species saturation. Several feeding niches within the plant seem to be open and new recruits are occasionally reported. Some become established (e.g. *Dectes texanus texanus*); others remain only as transient visitors despite sporadic colonizations (e.g. *Epilachna varivestis* in Illinois).

A few generalizations are possible at this time on the basis of these surveys and of a questionnaire filled out by 25 of the leading soybean entomologists in the producing states: 1) The recorded soybean arthropod fauna in the Nearctic region is composed of species recruited among the native fauna. There are no records of exotic pests of soybean in the region. 2) These species have been recruited from three main sources—a pool of polyphagous forb and grass feeders, a pool of oligophagous legume herbivores, and a small number of oligophagous non-legume associated herbivores. 3) None of the soybean insect pests have reached a dominant or key state in most regions. One exception may be *A. gemmatalis* in northern Florida. 4) Survey data from 21 of the 30 soybean producing states in the U.S. permit identification of several agroecological zones characterized by a communality of pest problems and faunal composition. 5) Under most situations feeding niches are still vacant and it is possible to identify exotic species in several parts of the world that have the potential to fill these niches. These species should be closely watched by federal and state quarantine authorities. 6) From the standpoint of ecological theory only niche analysis seems to hold promise in explaining arthropod community structure in soybean agroecosystems in the U.S. Neither island bigeography theory nor species/area analysis shed much light on the nature of extant communities.

ENTOMOPATHOGENS FOR CONTROL OF INSECT PESTS OF SOYBEANS

C. M. Ignoffo, SEA/AR, USDA, Columbia, Missouri 65205

Microorganisms which cause disease could be effectively and safely used to control insect pests of soybeans. Most of the insect pests have at least one pathogen attacking them and some pests have several. About 2 dozen pathogenic microorganisms have been isolated from insect pests of soybeans. Included in these isolations were 4 major viral types (nucleopolyhedrosis, cytoplasmic polyhedrosis, granulosis, entomopox) from at least 9 pest insects, 4 species of bacteria attacking 10 insect pests, 4 species of fungi attacking 12 insect pests and 7 species of protozoa attacking 11 insect pests. The possible use of three of these pathogenic microorganisms; a fungus (*Nomuraea rileyi*), a bacterium (*Bacillus thuringiensis*), and a virus (*Baculovirus heliothis*) in a season-long program against caterpillar pests of soybeans is presented and related to the concept of pest management and anticipated levels of control.

INSECT PEST MANAGEMENT IN NORTH CAROLINA SOYBEANS

J. W. Van Duyn and J. R. Bradley, Jr., North Carolina State University, Raleigh, North Carolina 27650

A system to prevent, tolerate, detect, and suppress insect pests of soybeans has been developed. The system is highly influenced by the corn earworm, *Heliothis zea,* and emphasizes cultural techniques to affect crop attractiveness and thresholds, promotion of indiginous natural enemies, scouting, and the selective use of insecticides. The management system and supporting research will be discussed.

Contributed Papers

PLANT, APHID VECTOR, AND PATHOGEN INTERACTION AS THEY AFFECT THE TRANSMISSION OF SOYBEAN MOSAIC VIRUS

J. A. Schultz, Department of Agricultural Entomology, University of Illinois, Urbana, Illinois 61801

Several factors involved in plant-vector-virus relationships were examined as to their effects on the epidemiology of SMV. Maximum field spread of SMV in experiments at the University of Illinois in 1978 from rows mechanically inoculated with virus to healthy rows occurred during the peak period of aphid abundance from August 6 to August 27. In field experiments with 2 'Clark' isolines possessing foliage of different colors, a low preference of alate aphids for light over dark green leaves as associated with a 36 percent reduction in the spread of SMV from inoculated to healthy rows in light green plots. Greenhouse studies showed that 'Clark' soybean plants of different ages were equally susceptible to infection with SMV by aphid inoculation. Aphid vectors recovered virus from flowers, green bean pods, mature, and young leaves from plants of all age groups 36 to 41 days after inoculation.

TEMPORAL PATTERNS OF TRANSIENT APHIDS IN SOYBEAN FIELDS AND THEIR IMPLICATIONS IN THE FIELD SPREAD OF SOYBEAN MOSAIC VIRUS

M. E. Irwin, R. M. Goodman, and G. A. Schultz, Departments of Agricultural Entomology and Plant Pathology, and the International Soybean Program (INTSOY), University of Illinois, Urbana, Illinois 61801

Soybean mosaic virus (SMV) occurs wherever soybeans are grown and, from a global perspective, is probably the most common and important virus attacking soybeans. The impact of SMV on soybean yield varies depending on locality, soybean cultivar, and particular strain of the virus. We have recently found large yield reductions and decreased seed quality in soybeans inoculated with a severe strain of SMV soon after emergence. Later inoculations produced less drastic results. SMV is transmitted from generation to generation through the seed and is spread within and between fields by transient alate aphids. The only known important primary field sources of SMV are soybean plants grown from infected seed. Timing and abundance of alate aphids are critical factors in the spread of SMV. More than 60 species of aphids land in central Illinois soybean fields each year. Since aphids do not colonize soybeans in the United States, insecticide sprays are ineffective against field spread of SMV. Only about six of the more abundant aphid species seem to be important as vectors of SMV in Illinois: *Aphis craccivora, A. citricola, Macrosiphum eurphorbiae, Myzus persicae, Rhopalosiphum maidis,* and *R. padi.* Flight activity patterns of all aphid species alighting in soybean fields in central Illinois have been monitored for the past four years. Results indicate that there is tremendous yearly variation in species mix, aphid abundance and flight timing. Because SMV reduces yields and lowers seed quality most severely when it infects soybean plants during early vegetative stages, late spring to early summer flights of aphid vectors are very important to SMV spread.

ALTERING FIELD SPREAD PATTERN OF SOYBEAN MOSAIC VIRUS BY MANIPULATING NATURAL BARRIERS

S. E. Halbert, M. E. Irwin, and R. M. Goodman, Departments of Agricultural Entomology and Plant Pathology, University of Illinois, and the International Soybean Program (INTSOY), Urbana, Illinois 61801

Soybean mosaic virus (SMV) is non-persistently transmitted by many species of aphids with varying degrees of efficiency. None of these species colonize soybeans, so SMV is transmitted within a field entirely by transient alatae. Sunflower barriers were tested for their effect on the spread pattern of SMV in soybean fields. Results showed that living barriers of sunflowers within a soybean field drastically altered the spread of the virus. Our data indicate that spread was substantially reduced for about 17 m downwind of the barrier. Transmission efficiency of live-trapped aphids on the leeward and windward sides of a sunflower barrier downwind of an SMV infected soybean source differed, with greater efficiency on the windward side of the barrier, although similar trapping positions where no barrier was present gave similar results. Catches from impact sticky traps within and above a sunflower barrier demonstrated that many aphids flew (or were carried) over the barrier rather than landed in it.

INSECT PESTS OF SOYBEANS IN BRAZIL: THEIR IMPORTANCE, DISTRIBUTION AND NATURAL ENEMIES

A. R. Panizzi, B. S. C. Ferreira and I. C. Corso, Centro Nacional de Pesquisa de Soja, EMBRAPA, Londrina, Parana, Brazil

Soybean production in Brazil has increased considerably during the last decade and a production of 14,400,000 tons is estimated for 1979, making Brazil the second largest soybean producer. With increasing production there has been a corresponding increase in problems of insect pests. Among the various species that attack soybeans, the major pests are stinkbugs, defoliating caterpillars and lepidopterous borers. The Hemiptera include more than 15 species, of which only three pentatomids [*Nezara viridula* (L.), *Piezodorus guildinii* (West.) and *Euschistus heros* (F.)] are economically important: these are the principal economic pests of soybean in Brazil. *N. viridula* is common in the South and rare in the Central West and further North. *P. guildinii* and *E. heros* are more abundant in the Central West although they do occur in all the soybean production areas. Defoliating caterpillars include several species of noctuids: *Anticarsia gemmatalis* Hubner and *Pseudoplusia includens* (Walker) are the most important. The widespread species *A. gemmatalis,* is the principal defoliator of soybeans and is more important in southern areas. *P. includens* is the second most important caterpillar, and is reported to cause great damage in the State of Sao Paulo. Borers include the shoot and axil borer *Epinotia aporema* (Walsingham) (Tortricidae), which occurs in all areas. The widespread stem borer, *Elasmopalpus lignosellus* (Zeller) (Phycitidae) is normally of less importance, although it causes considerable damage in first year of cultivation in the Cerrado region of Central Brazil.

In all areas the most important natural agent of control of *A. gemmatalis* and *P. includens* is the fungus *Nomuraea rileyi* (Farlow). It occurs earlier in the Central West moving to the South as the season progresses. Other pathogens include *Entomophthora sphaerosperma* (Fresenius), *Beauveria bassiana* (Balsamo), and a nuclear polyhedrosis virus, which is the second most important natural enemy of *A. gemmatalis*. *Nabis* spp., *Geocoris* spp., various Carabidae and spiders are the principal predators and they occur in all areas. There are several parasites (Hymenoptera and Diptera): the most important being *Microcharops bimaculata* (Ashmead) (Ichneumonidae) and *Litomastix truncatellus* (Dalman) (Encyrtidae) parasitizing *A. gemmatalis* and *P. includens*, respectively. The scelionids *Telenomus mormideae* (Costa Lima) and *Microphanurus scuticarinatus* (Costa Lima) are important egg parasites of *P. guildinii* and the tachinid *Eutrichopodopsis nitens* (Blanchard).

APHID VECTORS OF SOYBEAN MOSAIC VIRUS IN THE PHILIPPINES

D. R. A. Benigno and S. Boonarkka, UPLB, Laguna, Philippines

Eight aphid species commonly occurring in multi-cropped fields were tested for their potential as vectors of soybean mosaic virus. Only five aphid species, listed in descending order of efficiency, transmitted the virus: *Aphis glycines* Mats., *Myzus persicae* Sulz., *Aphis craccivora* Koch., *Hystereneura setariae* Thomas, and *Melanopsis indosacchari* David. The three other species tested, but failing to transmit the virus, were *aphis gossypii* Glover, *Longuinguis sacchari* Zehnt., and *Rhopalosiphum maidis* Fitch. The relationship of the virus and vectors is described.

MONITORING TRANSIENT APHID LANDING RATES FOR VIRUS EPIDEMIOLOGY STUDIES

M. E. Irwin and S. E. Halbert, Department of Agricultural Entomology and the International Soybean Program (INTSOY), University of Illinois and Illinois Natural History Survey, Urbana, Illinois 61810

Evaluation of field epidemiology of aphid-transmitted pathogens is in part dependent on monitoring alatae flight activity. Most methods currently in use rely on yellow pan traps or vertical sticky-coated impact traps. Trap design should bias catches to reflect absolute numbers on alate aphids landing/area/time interval. The purpose of this study was to test design variables and compare them with absolute landing records on plant foliage. A randomized complete block design of trap types was tested to detect the influence of the following variables on trap design: 1) orientation, horizontal vs. vertical; 2) landing surface, sticky vs. liquid; 3) height, canopy level vs. ½ m above canopy level; and 4) substrate color, clear glass vs. aluminum plaque vs. mirror vs. black tile vs. standard yellow plaque vs. ermine lime tile. Aphid landing rates were then recorded for the horizontally oriented ermine lime tile vs. a similar surface area of soybean foliage. Data indicate that vertical orientation collected statistically more aphids than did horizontal orientation, that aphid catches on sticky surfaces were not significantly different from those on liquid surfaces, that aphid catches on traps at canopy level were not significantly different from those ½ m above canopy level, and that aphid catches on differently colored traps were, black = aluminum = mirror = clear glass < ermine lime < standard yellow, differences significant at the 5 percent level. Aphid landing rates on horizontal ermine lime tiles were about equal to those on equal surface areas of soybean foliage. Horizontal ermine lime tile catches at or slightly above canopy level and either coated with a sticky substance or placed in a liquid-filled pan closely estimated the absolute landing rates of aphids on plant foliage. Yellow pans and vertically oriented traps overestimated absolute landing rates while black or u.v. reflective surfaces grossly underestimated absolute landing rates.

ANALYSIS OF INSECT POPULATION DYNAMICS IN SOYBEAN PRODUCTION

G. H. Smerage, Department of Agricultural Engineering, University of Florida, Gainesville, Florida 32611

Integrated crop-pest management is much more complex in concept and practice than conventional chemical methods. It virtually demands application of the analytical power of systems analysis. A major subsystem, the insect complex, is used in this paper to illustrate analyses that, ultimately, must be integrated with compatible analyses of the crop, other pest categories, and environmental subsystems of the soybean production ecosystem. During a crop season, a complex, which often changes between years, of significant pest and beneficial insect species becomes established in a soybean field and surrounding areas. Variations of within-field populations over the season arise from the interplay of numerous fundamental, biological and ecological processes. Principal objectives in using systems analysis as a tool for effective management of insect pests and crop damage are: 1) detailed identification and quantitative description of significant pest/beneficial species and processes in the composition of the insect complex; 2) comprehensive quantitative and qualitative knowledge about behavioral properties of the complex; 3) prediction of the temporal trajectories of populations from current levels in response to inherent factors, biotic and abiotic environmental stimuli, and human control; and 4) prediction of the nature and dynamics of accompanying crop damage.

A conceptual model of the insect population system is a network of fundamental, species and age-class specific, populational processes: storage, development, mortality, oviposition, predation, parasitism, competition, migration, and the complicity of crop, weather, and cultural factors. The processes and structure of the system are described by relations between relevant densities and flows of populations. A mathematical model describing the total system is formulated from the process and structural descriptions. Analysis of system behavior is performed on the mathematical system model by simulation and more mathematical methods. By simulation, temporal variations of populations over an interval are predicted as specific responses to the initial populations and the weather, migratory, and cultural inputs to the system over the interval. Mathematical analyses reveal more general behavioral features of the system—general behavioral modes, steady states, limit cycles, stability, and controlability—which are the basis for specific responses. Simple examples, using a mix of real and hypothetical data, illustrate modeling and behavior analysis by simulation for the soybean insect complex. Simulations show potential population responses to crop phenology, migration, insecticide application, and predator-prey interactions. These exercises facilitate understanding of knowledge extant, identification of voids to be filled, and the design of experiments to be performed.

EVALUATION OF SOYBEAN GENOTYPES FOR RESISTANCE TO STINK BUGS

D. F. Gilman, R. M. McPherson, C. Williams, and L. D. Newsom, Louisiana Agricultural Experiment Station, Baton Rouge, Louisiana

Several stink bug species, especially the Southern green stink bug (*Nezara viridula* L.), damage soybeans grown in many areas of the Southeastern United States. All adapted soybean cultivars are susceptible to stink bugs. Infested plants usually exhibit a delay in maturity and a reduction in both seed quality and yield. A study was initiated in 1975 to evaluate Maturity Groups V through VIII of the soybean germplasm collection for resistance to stink bugs. Lines were evaluated in replicated single-row plots subjected to high stink bug populations that developed on the early maturing cultivar Hill which was planted throughout the experimental material. These populations were supplemented by periodic releases of collected stink bugs. Damage ratings were made by separating seeds from each line into four categories (none, light, medium, and heavy) depending upon amount of visual damage. Results indicated that genetic variation for stink bug resistance was present in the germplasm collection, with damage ratings ranging from 0 to 100 percent of the seed with heavy damage. Several plant introductions had a moderate level of resistance, and one, PI 171444, exhibited a high level of resistance. These lines should be of value in breeding programs.

VARIETY SELECTION AND PLANTING DATE MANIPULATION AS CULTURAL CONTROLS FOR THE VELVETBEAN CATERPILLAR, *ANTICARSIA GEMMATALIS* HUBNER

D. C. Herzog, Department of Entomology and Nematology, University of Florida, Agricultural Research and Education Center, Quincy, Florida 32351

The velvetbean caterpillar, *Anticarsia gemmatalis* Hubner, is the most important and most widely distributed defoliating soybean insect pest in the Western Hemisphere. In the southeastern United States it is a late season pest, rarely developing to economically damaging levels before mid-August. This pest exhibits a decided preference for late-planted, late-maturing soybean varieties, and, hence, crop phenology is an important consideration in the protection of the crop from this pest. In 1977 a study was initiated to quantify the effect of variety (maturing grouping) and planting date on velvetbean caterpillar populations, defoliation resulting from their feeding and subsequent yield effects. In 1978, varieties from groups V (Forrest), VI (Centennial), VII (Bragg), and VIII (Hutton and Cobb) were planted at two-week intervals beginning May 18 for four dates of planting. A portion of each planting was protected from velvetbean caterpillar feeding throughout the season. Velvetbean caterpillar larval populations were monitored weekly and crop phenology twice weekly (Fehr et al. 1971) throughout the season.

Results showed that with delay in date of planting and with advance of maturity date, the crop supported larger populations of velvetbean caterpillar. However, examination of larval populations at comparable stages of crop phenology gives a more realistic representation of total impact on the crop. Later planted and later maturing varieties supported larger insect populations which occurred earlier in crop phenology and were sustained for much longer periods of time. During the 1978 cropping season all unprotected plantings suffered 100 percent defoliation. Only the May 18 planting of Forrest sustained no losses in yield. Delay in both planting and maturity resulted in a near linear relationship with increased losses in yield. In conclusion, one of the most effective methods of managing velvetbean caterpillar populations is through pest evasion, by early planting and/or selection of early maturing (short season) varieties. It is essential that this tool be incorporated into pest management programs. Further, it is recommended that a major breeding effort be directed at the development of high-yielding short season varieties.

SCREENING FOR INSECT RESISTANCE IN SOYBEAN BREEDING

S. C. Anand and J. L. Helm, McNair Seed Company, Laurinburg, North Carolina 28352

The soybean looper (*Pseudoplusia includens*) is becoming a major pest in the southern United States. As it is relatively difficult to control with available pesticides, it can cause severe losses. Today there is no commercial variety resistant to the looper. Thus, efforts are underway to transfer resistance from the unproductive PIs to agronomic cultivars. Because of difficulties in maintaining a uniform population of insects in the field and of the rearing of insects in the laboratory, no definite procedure has yet evolved for critical screening of soybean plants against the soybean looper. The present study involved the cross between resistant PI 229358 and 'Ransom,' a susceptible variety. The F_2 and F_3 generations were field grown in 1977 and 1978. Individual plants were screened in the laboratory against the soybean looper. Leaflets of the two kinds (three from each test plant and the susceptible check 'Lee 74') were stapled alternately on a paper in a circular pattern. The leaflets were edge trimmed to accommodate 6 of them on a 20 x 28 cm sheet of white paper. This paper with leaves was placed in a cake pan with a moist paper towel to assure a high R.H. Effort was made to use leaves of the same growth stage. Five loopers of approximately 2.5 cm length, collected from a soybean field, were released in each pan. The pan with covered and left overnight. In the morning the plants were scored resistant, moderately resistant, moderately susceptible and susceptible depending upon the extent of damage compared with the susceptible check in each pan.

Of the 112 F_2 plants screened, 17 were found to be resistant or moderately resistant. In the F_3 generation, 8 of the 17 F_2 resistant plants had progenies with more than 50 percent resistant or moderately resistant plants, while the remaining 9 progenies had more than half of the plants in moderately susceptible or susceptible categories. The observed data indicate to us that there are 3 major genes responsible for resistance to soybean looper in PI 229358. If the susceptible parent is designated aabbCC, the resistant parent will have the genetic constitution of AABBcc. The F_1 AaBbCc is susceptible. To be resistant, the plant has to have A_ and B_ with cc. In the F_2, a ratio of 54 susceptible to 9 resistant is expected. The observed ratio of 95 susceptible to 17 resistant was not far from expected ($X^2P = .75$). The resistant plants segregated 8 resistant to 9 susceptible progenies in the F_3 which was close to expected 9:7 ratio. The segregation within F_3 progenies indicates the likely presence of additional modifying factors. Working with the same resistant parent, Sisson et al. (1976) concluded the presence of 2 to 3 major genes responsible for resistance to Mexican bean beetle in PI 229358. It is very likely that the factor for resistance to Mexican bean beetle also imparts resistance to soybean looper. The technique described in this study could be effectively used for screening against soybean looper in a breeding program. The results of the back-cross generation would be needed to confirm these findings and this work is in progress in our research laboratory.

OBEREA BREVIS SWED. (COLEOPTERA: LAMIIDAE): A THREAT TO SOYBEAN CULTIVATION IN CENTRAL INDIA

O. P. Singh, Department of Entomology, Jawaharlal Nehru Agricultural University, Jabalpur, 482004 (M.P.), India

Soybean production in Central India is getting adversely affected due to increasing infestation by the girdle beetle *Oberea brevis* Swed. since 1969. The incidence, extent and nature of damage, host plants and biology of this pest were studied at Jabalpur (Central India). The pest is active from the end of July or the beginning of August until the crop matures. The infestation varies (from 13.4 percent in 1969 to 43.8 percent in 1976). It has a wide host range, causing considerable damage to soybean, green gram (*Phaseolus radiatus* L.), black gram (*P. mungo* L.) and eleven wild plants. The initial damage is caused by females by making two parallel girdles 15 cm apart on stems, petioles and placing eggs singly in the centrol holes made by them near the lower girdle. A female lays 8 to 72 eggs which hatch in 3 to 5 days. The infested twigs wither and dry. The larvae tunnel the stems and make them hollow. Most of the first generation larvae and nearly all the second generation ones overwinter and oversummer in 18 to 25 mm long pieces of stem, cut by the pest and plugged at both the terminals.

The losses caused by this pest have been assessed as follows: 1) When the pest attacks the seedling stage (15 to 17-day old crop) the plant mortality is high (up to 75 percent) and grain yield is reduced by 85.3 percent; 2) In the case of 1.5 to 2-month old plants, the plant mortality is negligible but the loss in grain yield is 66.7 percent; and 3) Additional loss in grain yield, caused by the hibernating and aestivating larvae by cutting the stems and lodging of the plants, has been 5.43 kg per hectare for every one percent infestation.

DEVELOPMENT OF SPIDERMITE RESISTANT SOYBEANS

B. H. Beard, USDA, SEA/AR, University of California, Davis, California 95616

Spidermites (*Tetranychus* species) only occasionally damage soybeans in the main production areas of the United States but can be extremely serious in hot dry areas. Spidermite resistant soybean germplasm has been developed cooperatively by the U.S. Department of Agriculture, Science and Education Administration–Agricultural Research and the University of California. At present there are nine resistant parents used in crosses with high yielding soybean belt cultivars. Segregates from these crosses have been tested for spidermite resistance in both the greenhouse and field. Greenhouse tests of spidermite resistance are time consuming, and only a limited number of lines can be tested at one time. However, greenhouse tests are more reliable and controllable. Field testing for spidermite resistance is subject to environmental conditions. Spidermites have reproduced at a slower rate on resistant lines compared to susceptible lines in both greenhouse and field tests.

EFFECT OF TILLAGE METHODS ON NEMATODE POPULATIONS AND THEIR VERTICAL DISTRIBUTION IN A SOYBEAN FIELD IN BRAZIL

P. S. Lehman and H. Antonio, Division of Plant Industry, Gainesville, Florida and Centro Nacional de Pesquisa de Soja, Londrina, Brazil

A study was conducted in the State of Parana in Brazil to evaluate if some tillage options acceptable to soybean growers could be integrated into a management system to control plant-parasitic nematodes.

Three types of cultural practices were used: discing, which is the practice normally used for soybeans in this region; no till or direct planting, which is being introduced as a means of erosion control; and disc-plowing, which is often used prior to planting wheat, but seldom used prior to planting soybean. The disc-plowing treatment consisted of plowing at two different times during the months before planting which normally have low precipitation. Each treatment was replicated five times. In each plot, soil samples were collected at 10 different sites and at each site at five depths: 0-5, 6-10, 11-15, 16-20, 21-25 cm. Samples were collected at five times to evaluate the initial effects the tillage-treatments had on nematode populations and the changes that occurred during the growing season. *Helicotylenchus dihystera* and *Meloidogyne javanica* were the predominate species of plant-parasitic nematodes present in the experimental field.

The initial vertical distribution of *H. dihystera* was similar in all plots before plowing. Greatest populations of *H. dihystera* were found at 0-5 cm. For the first 25 cm which were sampled, a linear decrease in numbers of this nematode per volume of soil was found with an increase in depth. Populations of *H. dihystera* were reduced with plowing or discing. This reduction was principally at depths of 0-10 cm, which is the vertical zone in which the root system of young soybeans would first develop. Six weeks after the first plowing, at 0-5 cm, the number of *H. dihystera* was more than five times higher in plots that were not plowed than in plots which were plowed. At planting, plots that were plowed two times or plots that were disced had much lower populations of *H. dihystera* at 0-10 cm than no-till plots. Although initial populations varied considerably between treatments, relative changes in the total root-knot nematode populations were influenced by tillage practices in a similar manner as was observed for the spiral nematode, *H. dihystera*. The no-till/plowing root-knot population ratio increased after each time of plowing indicating the detrimental effect of plowing as compared to no-till. Plowing also had a greater detrimental effect than discing. The detrimental effect of plowing on *M. javanica* populations as compared to no-till was consistently evident at 6-25 cm, but not at 0-5 cm.

Although plots that were plowed or disced had fewer plant-parasitic nematodes than no-till plots at planting, during the growing season the rate of increases on the susceptible cultivar Vicoja was greater in plots that had been plowed or disced than in no-till plots. Therefore, consideration should be given to the use of tillage practices that reduce populations initially and resistant cultivars that maintain populations at low levels during the growing season.

SEASONAL FIELD MAPPING OF SOYBEAN CYST NEMATODE POPULATIONS AS INFLUENCED BY CULTIVARS AND BLENDS

S. P. Caldwell and O. Myers, Jr., Southern Illinois University, Carbondale, Illinois 62901

Field populations of Race 3 soybean cyst nematode (*Heterodera glycines* Ichinohe) were monitored throughout the summer of 1978 on 5 resistant cultivars, 13 susceptible cultivars and 12 blends of 50 percent resistant and 50 percent susceptible cultivars. Soil samples were taken from the root zone within each row of three row plots in three replications. Twelve probes were taken to get a representative population from each plot. The first sample was taken before field preparation and four additional samples were taken at 25 day intervals after planting. Prepreparation cyst counts corresponded well with data obtained in 1977 when a final season count of similar plots in the same area ranged from 33 to 1133 cysts per 250 cc of soil. Cyst numbers of 70 per 250 cc of soil are generally regarded as sufficient to cause visual damage.

The first count following planting indicated that the population has been redistributed with cyst counts ranging from<33 to 233 with the majority of the plots less than 100. This redistribution was more uniform than expected and demonstrates preplant tillage practices may be important in cyst spread. Later counts showed that cyst numbers in susceptible plots increased rapidly with final counts of above 1000 in many plots. Cysts increased steadily but not as rapidly with final counts between 300 and 400. Cyst numbers on resistant cultivars remained low '(33 to 66 cyst/plot). The slower buildup of cyst populations and lower final cyst counts on blends as compared to susceptible cultivars suggest that blends might be useful in slowing or preventing a population shift to Race 4 of the soybean cyst nematode.

SOYBEAN ENTOMOLOGY INFORMATION SYSTEMS

J. Kogan and J. K. Bouseman, Section of Economic Entomology, Illinois Natural History Survey; Office of Agricultural Entomology, Illinois Agricultural Experiment Station; and INTSOY, Office of International Agriculture, University of Illinois, Urbana, Illinois 61801

The Soybean Insect Research Information Center (SIRIC) is a computerized information and retrieval system for the world-wide literature of arthropods associated with soybean. SIRIC is primarily a service oriented unit, operating in close cooperation with the Soybean Entomology Research Team of the Illinois Natural History Survey and the University of Illinois. SIRIC compiles, searches and retrieves printed documents on arthropods associated with soybean, including phytophagous species, vectors of soybean diseases, beneficial and incidental species which are part of the various soybean ecosystems around the world. Created in 1969 SIRIC was originally designed as a manually operated system capable of handling a few thousand citations. In 1969 there was only a small number of entomologists working full time on soybean pests in the United States. Currently there are more than a dozen different centers, especially in the Midwest and Southeast, with entomologists working in multi-disciplinary soybean research programs. A sharp increase in the numbers of published papers was a direct result of the research conducted in these institutions. With the expansion of the data base SIRIC developed a computerized system to better respond to requests and more efficiently handle this expanded literature.

SIRIC is a component of a broad soybean entomology information data base. The second component of this data base is the International Reference Collection of Soybean-Associated Arthropods (IRCSA) which stores actual specimens and ecological data from samples from all soybean producing regions of the world. IRCSA currently houses about 150,000 identified specimens of soybean-associated arthropods in over 2300 taxa. These specimens have been received from 23 soybean producing states in the United States and from around 35 foreign countries. The major foreign accessions have been from Latin America, with Brazil, Colombia and Mexico especially well-represented. Major emphasis is placed upon the quality of the identifications of the material in the collection. To assure a very high level of authoritativeness in the determinations, much material is submitted to outside systematists who are recognized experts in their fields for verification. To this end, a cooperating network of over 130 systematists in both domestic and foreign academic institutions and museums has been assembled. The purposes of IRCSA are fourfold, as follows: 1) To survey the soybean-associated arthropods of all soybean producing areas of the world with major emphasis on the phytophagous species and their parasites and predators, 2) To monitor the major pest species for possible changes in distribution, 3) To aid in predicting insect problems in areas initiating soybean production, and 4) To provide identification services for soybean researches and producers.

EFFECT OF OPEN AND CLOSED CANOPY ON SOYBEAN POD DAMAGE FROM *H. ZEA*

J. M. Joshi and A. Q. Sheikh, Soybean Research Institute, University of Maryland, Eastern Shore, Princess Anne, Maryland 21853

It has been previously reported that soybean fields with closed canopy escape pod damage from corn earworm (*H. Zea-Boddie*), but the recent reports from extension entomologists are controversial in this regard. This experiment was conducted to test the effectiveness of closed canopy in controlling pod damage. Variations in the closeness of canopy were achieved by using different seed rates (4, 8, and 12 seeds/0.3 m) and row spacings (11, 23, 46, and 91 cm wide). The soybean cultivar Delmar was planted on three different dates representing normal planting (May 13) and late planting (June 24 and July 8) during 1977 in a split plot design with four replications. The number of damaged pods were recorded on maturity in 4.9 m long rows. Pod damage was highest (28 pods) in the July 8 planting and lowest in the June 24 planting. The pod damage was highest for 8 seeds/0.3 m but was not significantly different from 12 seeds/0.3 m. Minimum pod damage was observed at 4 seeds/0.3 m. Pod damage was lowest in rows 11 cm wide but this damage was not significantly different from rows 46 cm wide. Rows 23 and 91 cm wide produced the same number of damaged pods (27 pods). The interaction between planting dates and seed rates, planting dates and row spacings and between seed rates and row spacings were highly significant. Based on our one year results, these data indicate that closed canopy may not help in reducing pod damage in every situation.

RESISTANCE TO *HELIOTHIS ARMIGERA* AND *HELIOTHIS PUNCTIGERA* IN THREE SOYBEAN LINES

L. D. Tuart and I. A. Rose, New South Wales Department of Agriculture, Narrabri N.S.W., Australia

Three soybean lines P.I. 171451, P.I. 227687 and P.I. 229358 have been reported to be resistant to various leaf-feeding insects including *Heliothis zea* and *Heliothis virescens.* A laboratory feeding trial was conducted comparing them to the commercial cultivar 'Bragg' for resistance to the two local *Heliothis* species. Thirty-six larvae of *H. punctigera* and 18 larvae of *H. armigera* were tested on each plant genotype. At day 11, 11, larval weights of *H. armigera* were 1277 mg (Bragg), 587 mg (P.I. 171451), 489 mg (P.I. 227687) and 876 mg (P.I. 229358). Corresponding weights for *H. punctigera* were 387, 249, 88 and 211 mg. For *H. armigera* total mortality at pupation was 28 percent (P.I. 171451), 56 percent (P.I. 227687), 39 percent (P.I. 229358) and 6 percent for Bragg. Time to pupation of surviving larvae was extended by 4 days when the three resistant lines were compared to Bragg. For *H. punctigera* mortality was 89 percent (P.I. 171451), 100 percent (P.I. 227687 and P.I. 229358) and 56 percent (Bragg). Time to pupation for larvae surviving on P.I. 171451 was extended by 4 days. The resistance of P.I. 227687 to both insect species was superior to the other lines which were superior to Bragg.

BEHAVIORAL AND NUTRITIONAL EFFECTS OF SOYBEAN PHYTOALEXINS ON PHYTOPHAGOUS INSECTS

S. V. Hart, M. Kogan, and J. D. Paxton, Departments of Entomology and Plant Pathology, University of Illinois, Urbana, Illinois 61801

Effects of soybean phytoalexins on the feeding of the adult Mexican bean beetle, *Epilachna varivestis* Mulsant (Coleoptera: Coccinellidae), and the larval soybean looper, *Pseudoplusia includens* (Walker) (Lepidoptera: Noctuidae) were investigated in the laboratory. Detached soybean cotyledons were inoculated with an extract of *Phytophthora megasperma* var. *sojae* to elicit phytoalexin production. Behavioral responses to the phytoalexins were analyzed using dual choice tests with uninoculated and inoculated cotyledons. Results from these tests indicate that both the Mexican bean beetle and soybean looper avoided feeding on the inoculated cotyledons containing phytoalexins. Additional studies are underway to ascertain possible effects of phytoalexins on nutrition of these phytophagous insects.

EFFECTS OF SIX LARVAL DENSITIES ON THE YIELDS OF THE SOYBEAN CULTIVAR JUPITER IN SOUTHERN TAMAULIPAS, MEXICO

S. de la Paz, Agric. Exp. Sta. "Las Huastecas", CIAGON-INIA-SIRH, Tampico, Tamaulipas, Mexico

During the 1977-78 soybean growing season in the southern area of the State of Tamaulipas, Mexico, an experiment was established at "Las Huastecas" and "Tancasneque" Agricultural Experiment Stations, CIAGON-INIA-SIRH. The purpose of this study was to determine the effect on soybean yield of several levels of larval density under field conditions. The levels were: 5, 10, 15, 20 (± 3) larvae ($>$ 1.5 cm)/row-m, and two checks (one insect free and one without chemical treatment). The main species present in both years were: *Anticarsia gemmatilis* (Hub.), *Trichoplusia ni* (Hub.) and *Pseuduplusia includens* (Walker). The peaks of defoliator infestation in both years were during the first 15 days of September (55 worms $>$1.5 cm/row-m), during the flowering period of the soybeans (R1, R2). In 1977, the rainfall during the pod-fill (R5, R6) was not as well distributed as in 1978.

Yields in the insect free plots were 2023 kg/ha and 2811 kg/ha for 1977 and 1978, respectively. Yields in the non-treated plots were 1312 kg/ha and 1783 kg/ha for 1977 and 1978, respectively, representing yield reductions of 25 and 36 percent for the check without chemical control.

In 1977 the insect free plot (13 applications) yielded 15 percent more ($P < .05$) than plots with 5, 10, 15, or 20 larvae/row-m. However, if the cost of the chemical applications are subtracted from the price of the soybeans, the net gains/ha were $359 for the insect free plot and $367 for the 20 worms/row m plot. In 1978, when the rainfall was well distributed during pod fill (R5-6), there were no significant differences between the insect free plot (8 applications) and the plot with 20 worms/row-m (3 applications). Yields were 2811 kg/ha and 2699 kg/ha, respectively.

UTILIZATION

Invited Papers—OILS

RAW MATERIAL AND SOYBEAN OIL QUALITY

T. L. Mounts, SEA-AR, USDA, Peoria, Illinois 61604

Preharvest factors of field and frost damage to soybean plants and post-harvest factors of handling, transportation, and storage greatly affect the quality of the extracted oil. Export shipments of soybeans were sampled at ports of embarkation and at destination ports. Beans were cleaned, cracked, decorticated and flaked and the oil was extracted in laboratory simulations of standard commercial procedures. Soybean oil extracted by a commercial processor from field-damaged beans was also obtained for characterization. Analysis of the oils included free fatty acid, iron and phosphatide content, color, peroxide value, chromatographic refining loss and oil degumming.

Significant differences were determined between oil from whole beans and that from split beans relative to free fatty acid and iron content. Formation of nonhydratable phosphatides, which are detrimental to oil quality, was determined to be an indicator of field damage to soybeans and to occur during export shipment of soybeans.

STORAGE, USE AND STABILITY OF SOYBEAN OIL AND ITS PRODUCTS

R. G. Krishnamurthy, Kraft, Inc., Research and Development, Glenview, Illinois 60025

The degree of freshness of the oil and its products at the consumer end is primarily dependent on the conditions of handling and storage of the same from the point of production of the crude oil. Various methods used in handling and storage of the oil and its products at all stages of processing as well as marketing will be reviewed. Keeping and performance quality vis-a-vis their chemical characteristics will be discussed. Techniques used in determining their quality and stability will be evaluated.

NUTRITIONAL ASPECTS OF SOYBEAN OIL UTILIZATION

E. A. Emken, SEA-AR, USDA, Peoria, Illinois 61604

Soybean oil has a tremendous nutritional impact on the American diet. It supplies almost 60 percent of the visible fats consumed in the United States, with partially hydrogenated soybean oil contributing about two-thirds of this total. Both soybean oil and partially hydrogenated soybean oil are well absorbed and are good sources of both Vitamin E and the polyunsaturated essential fatty acid, linoleic acid. In the processed oil, there are generally no problems with contaminants, toxins or bacterial growth. Some concern has been raised about the nutritional effect of fatty acid isomers formed during partial hydrogenation. These isomers are present also to a lesser extent in ruminant fats. Additional information has recently been reported that describes the distribution, accumulation, turnover and catabolism of isomeric fatty acids. In particular, human studies have confirmed some of the results observed in animal experiments and have found some apparent differences. Improved gas chromatographic techniques indicate low levels of *trans* and positional fatty acids in human tissues. No differences were found between subjects who had died from coronary and noncoronary causes. Conflicting results have been reported on the effects of partially hydrogenated soybean oil on atherosclerosis, cancer and diabetes. Recent recommendations however do not suggest modification of the diet to reduce the intake of partially hydrogenated soybean oil.

FOOD AND INDUSTRIAL USES OF SOYBEAN LECITHIN

B. F. Szuhaj, W. E. Prosise and F. J. Flider, Central Soya Company, Inc., Ft. Wayne, Indiana 46802

Commercial soybean lecithin is reviewed as a complex, versatile, and unexplored product of the soybean. Lecithin manufacturing, physical/chemical properties and functionality will be developed along with the product classes. Application areas for discussion include Food Processing and use in: cosmetics, pharmaceuticals, dietary supplements, coatings, plastics and rubber, glass and ceramics, paper and printing, and miscellaneous applications. The potential role of lecithin and further research possibilities in Food Processing are discussed.

INDUSTRIAL USES

K. T. Zilch, Emery Industries, Cincinnati, Ohio

The utilization of soybean oil as a raw material for the manufacture of industrial chemicals depends primarily upon its price relationship to other fats and oils having similar physical/chemical properties as well as the idiosyncrasies of its fatty acid constituents.

Because soybean oil is a vegetable triglyceride containing a high percentage of C_{18} chain length monobasic acids, which are over 60 percent polyunsatured monobasic acids, we find soybean oil, its unsaturated acids and derivatives therefrom being utilized in the manufacture of coatings, polyurethanes, antibiotics, polymeric acids, emulsifiers, lubricants, plasticizers, and so forth. These industrial uses will be discussed relative to their chemistry and functionality.

Invited Papers—PROTEIN

ANTI-NUTRITIONAL FACTORS AS DETERMINANTS OF SOYBEAN QUALITY

I. E. Liener, Department of Biochemistry, College of Biological Sciences, University of Minnesota, St. Paul, Minnesota 55108

Soybeans are known to contain a number of biologically active factors which have been shown to exert an adverse physiological response in animals. Some of these are readily inactivated by heat treatment, while others are heat stable. Among the heat-labile factors, the trypsin inhibitors have received the most study. Unless inactivated by heat, the trypsin inhibitors can interfere with the growth of animals by causing pancreatic hypertrophy and accentuating a deficiency of the sulfur-containing amino acids. From the limited evidence available, however, it is questionable whether the trypsin inhibitors are of any practical significance in human nutrition. Brief mention will also be made of such heat-sensitive components as hemagglutinins, goitrogens, and anti-vitamin factors which may possibly affect the nutritional quality of soybeans. Brief consideration will be given to a number of biologically active factors in soybeans which are not destroyed by heat. Those which, under certain conditions, may pose a problem include allergens, flatus-producing oligosaccharides, and lysinoalanine.

SOY PROTEINS—THEIR PRODUCTION AND PROPERTIES

M. F. Campbell, A. E. Staley Mfg. Co., Decatur, Illinois 62525

Edible soy protein is one of the world's least expensive and highest quality sources of protein. Historically, soybeans have been processed primarily for oil and for meal to be used as feeds. In recent years the use of soy protein by human food has been increasing. For maximum utilization of soy protein in food products, many different soy protein products have been developed. At least seven major categories of soy protein products that can be used as food ingredients are available from defatted soy flakes. The problem of selecting the best soy protein product to use in specific food application is often confusing. The basic soy proteins and their important characteristics will be discussed.

Product differences, such as composition, nutrition, flavor, cost and product form, will be briefly described to provide a better understanding of the soy protein products available to the food industry.

PROCESSED MEAT, BAKERY AND DAIRY APPLICATIONS OF SOY PROTEINS

M. S. Cole, Director of Research, Archer Daniels Midland Co., Decatur, Illinois

The major food industry applications of soy proteins are in processed meats and bakery products. Dairy products represent a potential area of application for various soy protein products. The use of soy proteins in these foods is based on their functional characteristics and the economies to be gained by their use. Functional properties contributed by soy proteins include fat emulsification and emulsion stabilization, water and fat binding viscosity, film formation and gelation. The relationship between these functional properties and specific product applications is reviewed.

ANALYSIS OF SOY IN MEAT PRODUCTS

A. C. Eldridge, SEA-AR, USDA, Peoria, Illinois 61604

The amount of soy added to meat products is difficult to determine if for no other reason than that soy proteins so closely resemble meat proteins in chemical composition. Methods investigated have included microscopy and histological staining, electrophoresis, immuno-electrophoresis, chromatographic separation of naturally occurring peptides, isolation of specific peptides after selective hydrolysis, and the analysis for minerals and characteristic carbohydrates. The merits and weaknesses of these approaches will be discussed along with new procedures involving fluorescence, $^{13}C^{12}C$ ratios and N-terminal amino acid sequences.

ORIENTAL SOYBEAN FOODS

D. Fukushima and H. Hashimoto, Kikkoman Foods, Inc., Walworth, Wisconsin 53184

Soybeans have been widely used as a source of the traditional foods in Oriental countries, which can be divided into two groups, namely, fermented and nonfermented foods. The main fermented soybean foods are soy sauce (shoyu in Japan, chiang-yu in China), fermented soy paste (miso in Japan, chiang in China), sufu, tempeh and natto. Chiang which originated in China some 2,500 years ago was introduced into Japan during the seventh century and transformed into the present Japanese shoyu and miso, which are now quite different from their Chinese counterparts. The shoyu fermentation consists of koji fermentation by *Aspergillus* species and the subsequent brine fermentation which contains lactic acid and alcoholic fermentations. The characteristic appetizing aroma observed in the Japanese style of soy sauce (shoyu) is derived through a special brine fermentation from the components of the wheat which constitutes about one-half of the materials. During the last two decades, the fermentation technology and engineering of shoyu and miso has made great progress in Japan.

Sufu (Chinese soybean cheese) is a cheese-like product originating in China in the fifth century. It is made through fermentation by *Mucor* or a related mold from soybean protein curd called "tofu" which is made by coagulating soy milk. Tempeh is one of the most important soybean foods, originating in Indonesia. It is a cake-like product made by fermenting soybeans with *Rhizopus*. Sufu and Tempeh are not made and consumed in Japan. On the other hand, natto is the fermented soybean protein product in Japan. It is a whole soybean product fermented by *Bacillus* species and was originated in a North Eastern part of Japan about 1,000 years ago. Natto is served with shoyu and mustard. Besides these traditional fermented foods, a new fermented soybean product appeared on the market recently in Japan. It is a soy milk drink fermented by lactic acid bacteria.

The main nonfermented soybean foods are tofu, kori-tofu and yuba which are made from soy milk; kinako and moyashi. Tofu (soy milk curd) is made by coagulating the heated soy milk with calcium sulfate. The texture of the white curd is soft, smooth, and properly elastic. There are several traditional tofu derivatives which are prepared by deep-fat frying or by baking of tofu. The unique process of aging the tofu curd in a frozen state for the easy dehydration of this gelatinous curd was originated in ancient Japan in the regions with severe, cold winters. The resulting dried curd product having a sponge-like texture is called "kori-tufu." Yuba is produced by heating soy milk at a temperature just below the boiling point in a flat pan, and drying the resulting coagulant film on the surface of the soy milk. Kori-tofu and yuba are served after cooking with seasonings or other ingredients. These traditional products which are made from soy milk have been eaten for centuries in Japan. Kinako is the powder of roasted soybeans. The flavor and the digestability of soybeans are improved by roasting and powdering. Kinako is easten by sprinkling with sugar on rice-cake or cooked rice. Moyashi is the sprout of the soybean or greenbean. Soybean sprouts are one of the most popular vegetables in Oriental countries.

Soybeans have been an important source of not only protein and fat, but also vitamins and flavor for Oriental people for thousands of years. The possible universal acceptance of these traditional foods is discussed.

FOODS FROM WHOLE SOYBEANS

A. I. Nelson, L. S. Wei and M. P. Steinberg, Department of Food Science, University of Illinois, Urbana, Illinois 61801

The soybean offers the most practical and economical answer to the present shortage of protein for human food, both in the U.S. and developing countries.

China and some other Asian countries have used whole soybeans for preparation of human food products since before the Christian Era. These products, especially the soy milk, are known for their beany, painty flavor. This objectionable flavor development has always discouraged soybean use as human food in the U.S. as well as most other parts of the world.

University of Illinois scientists have developed simple methods to inactivate the lipoxygenase enzyme which causes the off-flavor. The treated beans are bland in taste and have a pleasing, nutlike background flavor. The antinutritional properties in row soybeans are eliminated during inactivation of the lipoxygenase enzyme system which results in maximum protein quality in the final foods.

This processing concept has been applied to preparation of foods for home and village use. Starting with the dry raw soybean, procedures have been modified to reduce total cooking time to 30 minutes for most products in order to effect energy and time saving. This paper describes the technology involved in home or village preparation of soybean containing foods such as milk, weaning foods, breakfast foods and fried patties. Some of these products contain sufficient protein to be considered as meat substitutes. Also, these products are designed to utilize as the non-soy component, indigenous sources of calories. Finally, the concept includes addition of the soy component into local and highly acceptable recipes.

Contributed Papers

A RAPID HYDROLYSIS FOR QUANTITATION OF SOYBEAN METHIONINE AND LYSINE FOR BROILER FEED FORMULATION

C. T. Young, Department of Food Science, North Carolina State University, Raleigh, North Carolina 27650

With increase in prices of the raw ingredients for broiler feed formulation, the poultry industry needs a more efficient utilization of the ration proteins. Presently, rations are formulated on the basis of average amino acid compositional data. Amino acid composition can be influenced by variety, area of growth, storage conditions, processing and other factors, consequently an accurate analysis of protein in corn-soy based poultry rations is necessary if optimum broiler growth at a lower cost is to be attained. Too, lack of adequate storage facilities at feed mills requires a rapid amino acid analysis of feed ingredients before blending. Therefore the primary objective of this study was to develop a method for the rapid analysis for the amino acid composition of soybeans.

Soybean lots must be properly sampled and ground to 20 mesh particle size to give an adequate representation of the entire lot of soybeans. The soybean samples (50-100 mg depending upon analyzer used) are weighed in duplicate into Kimax culture tubes; then 10 ml of 6N HCl is added and mixed with a vortex mixer. A second 10 ml portion of 6N HCl is added. The tube is flushed with high purity nitrogen gas, capped tightly, and hydrolyzed in a forced air oven at 145 C for 2 hr. The hydrolyzed samples are cooled to room temperature and transferred to 100 ml beakers. The culture tube is then washed 3 times with 10 ml of citrate buffer (pH 2.2). To this hydrolyzate is added 2 to 10 ml aliquots of 6N NaOH allowing the mixture to cool between each addition. The pH, which should be at 7 or above, is now adjusted to 2.2 with constant stirring. The solution is quantitatively transferred to a 100 ml volumetric flask with pH 2.2 buffer and mixed. A portion is centrifuged on a Beckman Microfuge for 5 min at full speed. An appropriate aliquot is then analyzed for methionine and lysine. Samples may be held for several days at room temperature or frozen at 18 C or lower for storage up to one year before analysis.

With this method it is possible to sample, run the analysis, and return the results in approximately 8 hr. Certain areas of caution are in order to obtain accurate results. The type of tubes and volume of acid used in the system are very critical. Additionally, it is necessary to determine optimum hydrolysis time for each oven. These analyses can also be shortened by using a wrist action shaker during heating. It is extremely important that none of the humin (black material) be added with the sample to the ion exchange column since it irreversibly binds to the resin resulting in high pressure and poor resolution. Using a Beckman analyzer, methionine and lysine can be measured in about 30 min using a special column. In our laboratory we routinely analyze all of the amino acids (except tryptophan which is destroyed in acid hydrolysis) on our Durrum D-500 instrument in about 1 hr per sample. With accurate analyses of methionine and lysine, it is possible to formulate the broiler feed more accurately for optimum growth. This gives a substantial reduction in cost of the feed ingredients. This rapid hydrolysate system is adaptable to corn, the other major feed ingredient of broiler rations.

ADSORPTION OF OFF-FLAVOR COMPOUNDS ON SOY PROTEIN

T. G. Aspelund and L. A. Wilson, Department of Food Technology, Iowa State University, Ames, Iowa 50011

Soybeans are a significant protein source, but only small amounts are eaten directly by humans. Probably the most important reason for this, is the undesirable flavor of most soybeans. Numerous reports have been published on the flavor of soybeans, but most have been confined to identification of flavor compounds; alteration of the flavor by empirical processing techniques or flavor panel studies of specific products. Little information exists on the binding of soybean flavor compounds to soy protein or the factors affecting their release. The objective of this study was to determine the heats of adsorption, Gibbs free energy and entropy of adsorption. This thermodynamic data can then be utilized to determine the mode and strength of adsorption of off-flavor compounds onto soy protein.

A gas chromatographic technique was used to obtain the heats of adsorption of homologous series of alcohols, aldehydes, ketones and methyl esters onto soy protein. Five microliters of the headspace from each of the twenty-four compounds was injected onto a glass, gas chromotographic column packed with soy protein isolate Edi-Pro A. Triplicate samples were run at 80, 90 and 100 C and their retention times determined and corrected for dead time in the column. The resulting data was analyzed by statistical linear analysis.

It was found that the following compounds all had statistically significant heats of adsorption (ΔH) (99 percent confidence limit): 2-hexanone, 2-heptanone, 2-octanone, n-hexanal, n-heptanal, n-octanal, methyl pentanoate, methyl hexanoate, methyl heptanoate, methyl octanoate, n-butanol, n-pentanol, n-hexanol, n-heptanol, n-octanol, n-nonane and n-decane. The heats of adsorption increased with increasing carbon number. The hydrocarbons adsorbed the weakest onto soy protein with only nonspecific (van der Waal's forces) interactions (ΔH values of 0 to -8 kcal/mole). The aldehydes, ketones and methyl esters, as a class, were not statistically different from each other, but were significantly different from alcohols and hydrocarbons. Aldehydes, ketones and methyl esters adsorbed with nonspecific forces and a single hydrogen bond (ΔH values of -6 to -13 kcal/mole). Alcohols were found to adsorb the strongest interacting with both nonspecific and specific forces, probably, forming two hydrogen bonds (ΔH of -10 to -18 kcal/mole). The heat of adsorption data was further supported by the free energy of adsorption (ΔG) and entropy of adsorption (ΔS). The alcohols had larger negative ΔG's (-4 to -1 kcal/mole) which increased the temperature (indicative of a physical adsorption process). Likewise, the ΔS values for alcohols were large and negative (-40 to -25 cal/mole °K) indicating a loss of freedom (randomness) which would result by the binding of the alcohols to the soy protein.

The gas chromatographic method is a useful tool for determining the heats of adsorption of gases onto soy protein. Utilizing this data, predictions can be made for flavor adsorption in more complex systems. Likewise, the temperature requirements needed to remove specific off-flavors can be determined. Flavor chemists can also utilize this data in the flavoring of soy protein products.

INACTIVATION OF SOYBEAN LIPOXYGENASE *IN SITU* BY COMBINATIONS OF ETHANOL AND HEAT TREATMENTS

M. Borhan and H. Snyder, Department of Food Technology, Iowa State University, Ames, Iowa 50011

We investigated soaking of soybeans (Amsoy 71) in ethanolic solutions (15 to 60 percent) at warm temperatures (40 to 60 C) as a useful procedure to inactivate lipoxygenase and to maximize protein solubility. Such a procedure would be most useful in treating soybeans for manufacture of products such as soymilk or soy curd for which no further drying is necessary.

The soybeans were treated by soaking, freeze drying, size reduction, and lipid extraction. We obtained a crude soy extract (CSE) by extracting the defatted meal with water. The CSE was assayed for lipoxygenase activity and for nitrogen solubility (NSI). Also, we examined the ethanolic soaking solutions to learn what soybean constituents were being removed during soaking.

At 4 C ethanolic soaking can inactivate lipoxygenase in the CSE, but soaking in water after ethanolic soaking can reactivate lipoxygenase. This renaturation is not a problem when ethanolic soaking is done at 40 to 60 C. Of the several sets of conditions examined, 45 C, 15 percent ethanol and 12 hr of soaking inactivated lipoxygenase and retained a NSI of 67. Using these or similar conditions, we experimented with addition of specific chemicals, change of pH or change of ionic strength in the soaking solution. None of these modifications helped to destroy lipoxygenase or to maintain a high NSI. Surprisingly, increasing the pH to 8 or 9 did not result in a higher NSI compared to pH 7. Soaking soybeans at 45 C in 15 percent ethanol for 24 hr extracted 2 percent protein and 6 percent oligosaccharides based on the original weight of soybeans. Of the total phytate, approximately 20 percent was extracted. These changes (loss of oligosaccharides and phytate) are beneficial for products consumed by humans.

To explore the temperatures and percent of ethanol required for inactivating lipoxygenase in minimal time, a series of experiments were done at 40 to 60 C and 15 to 60 percent ethanol. Data will be presented on the results of these experiments; results which will allow a processor to choose the time, temperature and percentage ethanol that would be most economical and still fit constraints for lipoxygenase inactivation and maximal NSI.

INTERESTERIFICATION OF SOYBEAN OIL WITH EDIBLE TALLOW

A. P. Handel and R. G. Arnold, Department of Food Science and Technology, University of Nebraska, Lincoln, Nebraska 68583

Interesterification is a relatively simple chemical process which results in the redistribution of fatty acids among triacylglycerol molecules. The resultant product has modified physical and chemical properties. This process has been used for many years to produce lard with a smooth texture.

The most prevalent technique used to modify the physical properties of soybean oil has been hydrogenation. Although this procedure results in a product with excellent physical properties and stability, there are several disadvantages to this procedure. During hydrogenation much of the desirable polyunsaturated fatty acid content is destroyed either being converted to saturated fatty acids or fatty acids with *trans*- rather than *cis*- double bonds. Since interesterification does not affect the double bonds of fatty acids, their nutritive value would not be changed. The simpler process and milder conditions of interesterification would also provide an energy savings over the high temperature and pressure hydrogenation process. It is our objective to produce a semi-solid plastic fat suitable for margarine and shortening manufacture by interesterifying soybean oil with edible tallow.

Work in Eastern Europe and at the Northern Regional Research Laboratory in Peoria have shown that a semi-solid product could be produced by interesterification of a vegetable oil with solid fats. In Bulgaria, sunflower oil was interesterified with lard and tallow. In Peoria, soybean oil was interesterified with fully hydrogenated soybean oil to produce a product with no *trans*- fatty acids. Our procedure involves the mixing of a dried oil with 0.5 percent sodium methoxide catalyst and stirring for 30 minutes at 70 C. The catalyst is then deactivated with water and the oil washed with water and then dried. The extent of interesterification was determined by lipase hydrolysis and analysis of the resulting 2-monoacylglycerols. Physical properties of various mixtures were compared using each mixture's solid fat index. Stability of the products was also determined.

Our results show that a semi-solid fat product can be produced by interesterification of soybean oil and tallow. Varying physical properties can be obtained by varying the proportions of soybean oil and tallow and these properties can be predicted from solid faty index data of known mixtures. Stability data on these products will also be presented. Interesterification seems to be a valid alternative for the preparation of semi-solid fat products. Soybean oil has been interesterified with tallow to produce a product with desirable physical properties, stability and nutritive value.

EFFECT OF FORMULATION AND PROCESSING VARIABLES ON THE QUALITY OF SOYBEAN YOGURT

J. K. Orlowski, A. I. Nelson, and L. S. Wei, Department of Food Science, University of Illinois, Urbana, Illinois 61801

Vegetable proteins, especially soybean, hold promise for future human food use. Soybean yogurt is such a product. Soybean yogurt was prepared using Illinois process soybean base developed by Nelson, et al., (1975). This base contains 7.5 percent soy solids, about 3.8 percent protein and is excellent for making yogurt since it is bland and has no beany off-flavors. A frozen culture of *Streptococcus thermophilus* and *Lactobacillus bulgaricus* was used to inoculate sterile, 2 percent fat cows' milk for preparation of a medium used for inoculation of soy base.

Soybean base samples containing 4, 6 or 8 percent of sucrose, cerelose or sucrose and cerelose in combination along with a control, no sugar added, were inoculated with the standard culture. The pH decrease of all samples was rapid initially and leveled off after 2½ hr. At all levels of sugar addition, the cerelose and sucrose:cerelose samples fermented to the lowest pH values. These were followed by the sucrose sample and the no added sugar control. The final pH of the cows' milk control was similar to the cerelose and sucrose:cerelose samples. As sugar level increased from 4 to 8 percent, final pH values for each sugar decreased slightly; howver, no significant difference was observed between any added sugar sample. These results indicate the ability of *S. thermophilus* and *L. bulgaricus* to use the added sugars. In similar experiments, total titratible acidity increased sharply at first but leveled off after about 2½ hr of incubation. Sucrose:cerelose samples developed the greatest amount of titratible acidity followed closely by cerelose, then sucrose. The total titratible acidity of the no added sugar control was significantly lower and the cows' milk yogurt significantly higher than any other soybean sample.

Log plots of viscosity vs time showed about a 200-fold increase in viscosity during the first three hr. of incubation in most samples. No trend in increasing viscosity was observed with the same sugar at increasing sugar level. The no added sugar sample was lowest and cows' milk yogurt highest in viscosity. Both pH and total titratible acidity plotted against viscoity gave similar results and illustrated the dependence of viscosity on pH and acidity. At pH values in the isoelectric range a 125-fold increase in viscosity occurred for all soybean samples. No obvious trend in syneresis and curd strength with respect to formulation was observed. Samples for organoleptic evaluation were received favorably and were preferred over cows' milk yogurt by some testers. Six percent added sugar as 3 percent sucrose and 3 percent cerelose, fermented to a pH of 4.21, was the preferred sample. This was probably due to its desirable sugar/acid balance. The Illinois process soybean yogurt is also adaptable to flavoring with fruit and appears promising as a new food product.

EFFECT OF FREEZE DAMAGE ON SOYBEAN QUALITY AND STORAGE STABILITY

G. E. Urbanski, L. S. Wei and A. I. Nelson, Department of Food Science, University of Illinois, Urbana, Illinois 61810

When growth of the soybean plant is arrested prematurely by freeze damage, soybeans that would be yellow if matured normally, are green when harvested. Many authors have reported on the oil and meal quality of freeze damaged soybeans. However, the objectives of this research were to study the whole bean quality and storage stability of free damaged soybeans. These two factors, which have received little attention in the literature, are important to the understanding of how freeze damaged soybeans can be utilized.

Freeze damaged samples were prepared by harvesting green immature soybean pods from growing plants and freezing them at 22 F for six hours. After freezing, the pods were dried with room temperature air and the beans were removed from their pods. All samples were stored at room temperature. Samples were analyzed after preparation and at the end of 2, 4, 6, 8, and 14 months storage for protein, oil trypsin inhibitor activity, free fatty acid content of oil, photometric index of oil and lipoxygenase activity. Samples were organoleptically evaluated after preparation and after six months storage for color, flavor, off-flavor and texture. Undamaged soybeans of the same varieties as the freeze damaged soybeans were used as control samples.

The average oil and protein content remained fairly constant during storage. During 14 months storage, average trypsin inhibitor activity of the freeze damaged samples increased from 23.3 to 27.4, indicating a decrease in protein digestibility upon storage. Before storage, the average lipxoygenase activity (L.A.) of the freeze damaged samples (13.2) was lower than the average L.A. of the control samples (30.1), indicating that freeze damaged soybeans have a lower potential for development of off-flavors. During storage, the L.A. of all samples remained relatively stable. Storage resulted in an increase in average free fatty acid content (F.F.A.) from 0.26 to 1.63 percent for the freeze damaged samples and from 0.14 to 0.48 percent for the control. The average F.F.A. of severely damaged portions of the freeze damaged samples increased from 0.32 to 2.00 percent, while mildly damaged portions increased from 0.16 to 1.10 percent. Therefore, storage of freeze damaged soybeans greatly increases the losses associated with refining crude oil from them. Organoleptic evaluation of the cooked samples before storage showed freeze damaged samples inferior to control samples in color, flavor, and off-flavor. Evaluations of samples stored six months showed little change. The freeze damaged soybeans are of low whole bean quality due to poor flavor and increased off-flavor and photometric index. After storage, higher trypsin inhibitor and free fatty acid of oil lower freeze damaged soybean value for oil processing and for feed.

QUALITY ASSESSMENT OF AUSTRALIAN SOYBEAN VARIETIES FOR THE PRODUCTION OF JAPANESE FOODS

O. G. Carter, G. R. Skurray, J. Cunich and S. Honey, Hawkesbury Agricultural College, Richmond, N.S.W. 2753, Australia

Japan imports 3,600,000 M.T. of soybeans each year for food and crushing purposes. Tofu (bean curd) is the most important food product made from soybeans in Japan. Soybean varieties vary greatly in their chemical composition and the studies reported here were concerned with the texture, color, flavor and nutritional value of different varieties of soybeans.

Previous studies had shown that the ratio of 7S and 11S proteins and the phosphorus to nitrogen ratio was important in determining the texture of tofu. The present studies indicate that these ratios are not as important as the amount of calcium sulphate used in the preparation of tofu. An Instron food texture instrument (1140) and sensory evaluation were used. Variation in the ratio of 7S and 11S protein and protein to nitrogen ratio could be counteracted by changing the amount of calcium sulphate used to precipitate the bean curd. A standard laboratory procedure for tofu preparation was developed to rapidly screen twenty different varieties of soybeans and the method correlated significantly with tofu prepared in the traditional way. The nutritional value of soybean protein was determined by measurements of protein efficiency ratios and biological value. The protein in several varieties had equal nutritional value, however there was a marked variation between some varieties. The food and nutrition qualities of the different varieties have been compared with their dry matter and protein yield per hectare.

PROCESSING AND APPLICATION OF SOYA BEANS FOR HUMAN NUTRITION AT HOME AND VILLAGE LEVELS

T. G. Loo, Royal Tropical Institute, Amsterdam, The Netherlands

A considerable amount of high class research work has been carried out in Europe, U.S.A., and other parts of the world on the processing of soya beans. But almost all of this work has been focused (and is still being focused) on the more sophisticated soya products such as soya protein isolates and textured vegetable proteins, which can only be manufactured in relatively modern and expensive factories, on a large scale. Consequently, the results obtained are only of benefit for the developed countries with high standards of living. The less developed and developing countries generally cannot afford these soya products in their daily menu, because of the relatively high prices. Furthermore, it is relatively difficult to use them in their national diets. These nations would generally be obliged to adapt new food habits, if they want to make use of the results of the above-mentioned research.

It is of utmost importance that the less developed and the developing countries avail themselves of simple methods for processing soya beans into easily digestible soya products, which can be manufactured at home and village levels. These products should have a high nutritive value and preferably a neutral, bland taste, so that they can be incorporated into local dishes of the countries concerned, without the risk of being refused by the population. Because of the neutral bland taste, these soya products will conceive the taste and flavor of the national dish into which they are incorporated. In this way they will in most cases be directly accepted by the consumers. Soya products like soya milk and soya bean curd (also called toufu), which have proved to be high quality protein sources and which have existed for thousands of years, can easily be flavored or spiced so that they will become as delicious as meat and fish dishes.

Soya milk can be prepared easily at home and village levels just like soya bean curd and soya steak (also called tempeh). The most important factor is the absence of the beany taste in these products, and also the absence of trypsin inhibitors and of hemagglutinins. Especially for babies and toddlers these two antinutritive factors should be inactivated. As they both belong to the proteins, simple heat treatment during a few minutes at temperatures above 90 C is sufficient. With soya products prepared at home and village levels, a great variety of delicious dishes and high quality weaning food can be prepared.

USE OF WHOLE SOYBEANS TO IMPROVE PROTEIN LEVEL AND QUALITY OF CORN-BASED PRODUCTS FOR DEVELOPING SOCIETIES

J. L. Collins and J. F. Sanchez, Department of Food Technology and Science, University of Tennessee, Knoxville, Tennessee 37916

Corn is a staple foodstuff for millions of people in Latin America and parts of Africa. In Bolivia, corn is the most important cereal crop and is the most popular single ingredient in most dishes. The tortilla is a popular corn-based food in Bolivia and certain other countries. Where corn is the principal cereal, protein-calorie malnutrition may be found. The soybean is recognized as a protein supplement and may be used to fortify corn-based food. This study was conducted to: determine the effect of soy meal (soy) on quality of tortillas; evaluate the proteins in mixtures of corn, soy, and cheese for protein efficiency ratio (PER); and establish chemical scores for limiting essential amino acids (EAA) in the samples.

Tortillas were prepared with cornmeal, soy Monterey Jack cheese, salt, fat, and water. Soft, yellow kerneled corn and Forrest soybeans were cooked in water, dried, and ground into meal. Tortilla dough was prepared in eight treatments: four contained cornmeal and soy at 0, 10, 20, and 30 percent; four contained cornmeal, 10 percent cheese and 0, 10, 20, and 30 percent soy. Doughs were prepared to a consistency similar to that of the control (cornmeal only) by varying the amount of water. Tortillas were cooked in an iron skillet at 307 C for 4 minutes.

Firmness of tortillas decreased with increased soy and with cheese. Soy produced dough and tortillas which were slightly lighter in color than the control. A taste panel of Latin American students showed no preference for tortillas containing the different levels of soy; however, the panel did prefer samples with cheese. Mean flavor score was 5.7 on an 8-point scale (8 = like extremely). Samples with up to 20 percent soy were preferred as much as the control for mouthfeel; score = 5.4.

The cornmeal was deficient in the S-amino acids, lysine, tryptophan, isoleucine, and valine; soy and cheese were deficient in S-amino acids when compared to the FAO pattern. All mixtures of corn, soy, and cheese were deficient in the S-amino acids; other EAA were sufficient. Chemical scores were calculated using whole egg protein as the reference. PER of a rat diet containing corn as the only protein source was 0.87 against casein (2.5). When soy was raised to 10, 20, and 30 percent, PER was increased to 1.98, 2.36, and 2.48, respectively. Diets with 10 percent cheese, soy at 10, 20, and 30 percent, and cornmeal to raise the amount to 100 percent PER's which were not different from that of casein. Nutrition of people who consume primarily corn could be raised substantially by adding at least 10 percent each of soy and cheese to tortillas. Addition of only 10 percent soy would be a step in the right direction. This low level of soy should cause no serious organoleptic problem since the panel showed no preference among tortillas containing soy up to 20 percent.

UTILIZATION OF SOYBEAN [*GLYCINE MAX* (L.) MERR.] AS A SUPPLEMENTARY HUMAN FOOD TO COMBAT PROTEIN MALNUTRITION IN WEST BENGAL, INDIA

M. Roquib, Faculty of Agriculture, Bidhan Chandra Krishi Viswa Vidyalaya, West Bengal, India

An extensive measure has recently been taken up to extend soybean cultivation throughout the State of West Bengal, India. Simultaneously, work has also been started to popularize soybean as a food amongst the low to middle income group of people. In this context, an experiment was done during 1976-78: 1) to test the comparative acceptability of consumers (orphan boys and poorly paid workers) for incorporating soybean in the form of full-fat flour, whole soybean or soybean milk; 2) to test the reactions of consumers to *chapatties* (hand made bread) made from regular wheat flour to that made from different blends of these two flours as well as to quantify the most acceptable proportions; 3) to explore the possibilities of introducing soy-based tiffin foods and working out their cost; and 4) to impress upon the consumers the nutritional importance of soybean to increase its home consumption.

The preparation of four common foods using full-fat soyflour, and the flour using soybean grain, and eight others using soymilk or soymilk residue have been standardized. These can be followed with available resources of an average Bengalee village family. A follow-up study was then made with the soy-based foods prepared at different months of the year.

The study reveals that consumers have generally accepted the foods substituted with full-fat soyflour or with whole soybean. Very poor response could be noticed in accepting soymilk directly. However, the response was high in accepting food prepared from soymilk or soymilk residue. *Chappatties* made with 0, 10, 15 and 20 percent full-fat soyflour were equally acceptable in terms of texture, smell, taste and cooking qualities. With fresh and good quality soyflour, its proportion can be raised up to 25 percent in such foods. Soybean preparations were liked by people of all age groups and appeared to be cheaper in price as compared to that of other standard foods of similar kinds. As such, these foods can be incorporated in the daily diet to enrich the nutrition level of cereal food. Soybean, as a food crop, offers an important scope for fruitful exploitation to combat protein-calorie malnutrition in this region as well.

INTRODUCTION OF LOW COST EXTRUDED FOODS IN INDIA

R. N. Trikha and R. W. Nave, Soya Production and Research Association, Bareilly U.P., India

In a predominately vegetarian society like India, fats and proteins of vegetable origin acquire special significance. Importance of soybean in Indian agriculture, therefore, is obvious. Soybeans were primarily used as pulses by the local population and the green and dried vegetative parts were used as forage for cattle. Among factors responsible for the failure of soybeans to attract Indian farmers, the major ones were probably the nonavailability of well adapted, high yielding varieties, lack of knowledge of utilization, processing and inappropriate procedures and a poorly developed marketing system for soybeans. In view of the chronic shortage of protein foods and vegetable oil in the country, a fresh attempt to popularize soybeans in India was made in 1965.

Unfortunately, no scheme in the field existed to coordinate farmers in the production, marketing, processing and utilization. Efforts were therefore made by the Soya Production and Research Association in collaboration with G. B. Pant University of Agriculture and Technology to undertake an ambitious intensive soybean project having an integrated approach to research, extension, processing, utilization and popularization. The Association imported and installed the first extrusion cooker x-25 followed by x-155 from Wengers, U.S.A., and started manufacture of extrusion cooked texturized vegetable protein foods based on defatted flakes and full fat soya flour. Several combinations have been tried with corn, wheat and rice.

Currently, a considerable portion of soybean produced in India is being solvent extracted. The cake is exported for feed and the oil is used for hydrogenated fat production. A substantial quantity of full fat and defatted soy flour is used in feeding programs, bakeries and as cattle feed. The properly processed and favorably priced soy protein foods have a good potential market in India. It is hoped that this production of low cost extrusion foods will help in expansion of soybean cultivation and will permit India to combat protein calorie malnutrition in India by 1985.

BREEDING

Invited Papers

ROLE OF CLASSICAL BREEDING PROCEDURE IN IMPROVEMENT OF SELF-POLLINATED CROPS

V. A. Johnson and J. W. Schmidt, USDA, SEA/AR, North Central Region and University of Nebraska, Lincoln, Nebraska

Many variations of classical breeding procedures have been employed for improvement of self-pollinating crop species. They are dictated mainly by the crop species involved, the production environment, breeding objectives and priorities and financial constraints. The breeding system followed in the cooperative USDA-Nebraska winter wheat improvement program is discussed. The system is based on the following assumptions: 1) Useful genetic variability in common wheat (*Triticum aestivum* L.) has not been exhausted. 2) Genetic variance for yield in common wheat is mostly additive and can be fixed in true breeding lines. 3) Significant increases in yield can result from the control or removal of such yield constraints as diseases and insects. 4) Progeny performance cannot be predicted with certainty from known attributes of the parent varieties. 5) Varietal heterogeneity can serve as a buffer against variations in production environment and can effectively contribute to performance stability. 6) Performance stability is as important as high yield potential in the hard winter wheat region of the United States. 7) Experimental wheats that remain in the breeder's nursery contribute little to agriculture. 8) New improved varieties should be perceived as transitory. They should be released with the expectation of replacement with better varieties in a few years. and 9) Genetic vulnerability constitutes a continuing potential threat to production the U.S. high plains where wheat dominates the crop acreage. Availability and use by growers of many varieties, relatively rapid turnover of varieties, and intra-varietal heterogeneity can reduce genetic vulnerability.

Three to four hundred crosses are made annually. Experience has shown that this number can be managed adequately with available land, facilities and resources. Hybrid populations normally are advanced as bulks through the F_3generation at which time they are head selected. Yield testing at 3 to 5 Nebraska sites is initiated in the F_6 generation with regional evaluation, farm testing and large-scale milling and baking trials, and foundation seed increase occuring in the F_9 and F_{10} generations. Of 20 hard red winter wheat varieties released for commercial production by Nebraska since 1963, 3 are increases of an F_2 plant, 15 were selected in the F_3 and 2 in the F_4 generation. Grower and industry acceptance of the varieties has been excellent. The "Scout" variety, an F_3 selection which exhibits heterogeneity in several agronomic traits, is adapted widely in the hard winter wheat region. Scout was released in 1963. By 1968, it had become the most widely grown wheat variety in the United States occupying more than 7 million acres. Nebraska varieties comprise more than 20 percent of the total U.S. wheat acreage. The productivity and performance stability of the Nebraska varieties support strongly early generation selection as an effective breeding procedure in wheat. We believe the heterogeneity that exists to some degree in each of the varieties contributes to their superior performance. It has not affected grower and industry acceptance of the varieties.

POPULATION IMPROVEMENT IN SELF-POLLINATED CROPS

D. F. Matzinger and E. A. Wernsman, Departments of Genetics and Crop Science, North Carolina State University, Raleigh, North Carolina 27650

Traditional breeding programs in the naturally self-fertilizing species differ considerably from those utilized with the naturally cross-fertilizing species. Most population improvement programs in self-fertilizing species have relied on the accumulation of many improve homozygous genotypes within a heterogeneous population, either by natural or artificial selection.

The present paper will outline procedures for controlled recurrent selection in the naturally self-fertilizing species, *Nicotiana tabacum*. The objective of these studies is to provide improved heterozygous populations from which to isolate superior homozygous genotypes. Results will be presented to demonstrate selection response for the primary character under selection, correlated response of other characters, and the efficacy of selection indexes to select in opposition to antagonistic genetic correlations.

LONG- AND SHORT-TERM RECURRENT SELECTION IN FINITE POPULATIONS

J. O. Rawlings, Department of Statistics, North Carolina State University, Raleigh, North Carolina 27650

The design of recurrent selection programs depends on the relative emphasis given to long-term and short-term objectives. Maximizing short-term selection gains requires high selection intensity with little concern given to the retention of genetic variance. Maximizing long-term selection progress, on the other hand, is dependent upon the conservation of genetic variability, the avoidance of genetic drift. This paper is a review of certain relevant literature concerning the design of recurrent selection programs. Particular attention will be given to the choice of population size so as to satisfy (nearly) both long-term and short-term objectives.

STRATEGIES FOR INTROGRESSING EXOTIC GERMPLASM IN SOYBEAN BREEDING PROGRAMS

W. J. Kenworthy, University of Maryland, College Park, Maryland

Utilization of exotic germplasm for improvement of seed yield has received little emphasis in soybean breeding. The large number of unadapted accessions in the germplasm collection has generally discouraged extensive yield evaluations of this material. Cooperative regional yield evaluations of the germplasm collection offer promise in identifying productive lines. Alternative breeding procedures for incorporating productive exotic germplasm into breeding programs will be outlined. A recurrent selection program conducted in a population consisting of 25 percent exotic germplasm will be described. The favorable yield response observed in this recurrent selection program suggests that the use of population improvement techniques may be beneficial in developing populations of greater diversity and productivity.

PHYSIOLOGICAL TRAITS AND PLANT BREEDING

D. N. Moss, Department of Crop Science, Oregon State University, Corvallis, Oregon 97331

Progress in plant breeding has often been rapid when the breeding objective is defined clearly and there are simple tests available to measure progress toward an objective. An example is breeding for disease resistance where the disease is controlled by a single gene, the pathogen can be cultured, and the test plants can be innoculated to assure the presence of the disease. In sharp contrast, breeding for high yield in the absence of major disease, insect, or environmental limitations most often is a trial and error process. If, in reality, physiology has a significant role to play in plant breeding its greatest promise lies in defining specific physiological processes that are limiting yields and providing testing procedures that will permit plants to be screened quickly for the "level" of the process in that plant, thereby enabling the breeders to search for genetic variability for the trait. Breeding for nitrogen responsiveness in wheat and rice is an example of where this principle has been applied with remarkable success.

What is the approach where limiting physiological processes have not been identified? If an understanding of yield limitations is to be reached, physiological research programs closely tied to variety development programs appear to offer the greatest opportunity of success. Plant design research on barley at the University of Minnesota illustrates the process, problems and potential of joint physiology-plant breeding research.

RESEARCH PRIORITIES IN PLANT BREEDING

G. F. Sprague, University of Illinois, Urbana, Illinois 61801

Progress in plant breeding is dependent upon two factors: the amount of genetic variability available and the efficiency of the selection methods employed. Genetic variability may be increased through use of a wider array of adapted material, through incorporation of unadapted exotic material, use of mutagenic agents, or possibly eventually protoplast fusion. The initial consequences of each of these methods of expanding genetic variability is a reduction in mean performance thus complicating the subsequent process of isolating improved cultivars.

Breeding procedures fall into two general classes; pedigree selection or recurrent selection. The efficiency under either approach depends upon the efficiency of screening techniques employed. Where quantitative traits are involved, recurrent selection appears to offer most promise. The development and utilization of new evaluation techniques has been an important factor in plant breeding programs. If genetic engineering is to play the role in plant breeding that some have claimed for it, it will be through providing new and simple evaluation techniques which are compatible with plant breeding operations.

Contributed Papers

DEVELOPING SOYBEAN VARIETIES FOR THAILAND THROUGH DISRUPTIVE SEASONAL AND LOCATIONAL SELECTION

S. Shanmugasundaram and A. NaLampang, Asian Vegetable Research and Development Center, Shanhua, Tainan 741, Taiwan, R.O.C. and Department of Agriculture, Bangkhen-9, Bangkok, Thailand

Soybean, *Glycine max* (L.) Merr. is grown during the rainy season (May-June to Aug.-Sept.) and the dry season (Dec.-Jan. to March-April) in Thailand. The rainy season crop is concentrated in the Central Plain, while the dry season crop is grown in northern Thailand under no-tillage rice-stubble culture. Crosses were made at the Asian Vegetable Research and Development Center (AVRDC) during 1973. In 1974, each bulked F_4 population from 120 crosses were evaluated at Chainat, Thailand, during the dry season. Selected F_5 lines from 71 crosses were screened at AVRDC. During the dry season in 1976, F_6 was evaluated with two replications at ChiengMai, Thailand. At the Praputtabat Agricultural Experiment Station during the rainy season 80 F_7 lines were compared with local varieties in a regional trial, with four replications. In a second regional trial planted at Chainat, ChiengMai, and Pitsanulok during 1977-78, the genotype, season, location, year and the first order interactions were investigated.

Through visual selection, 422 single plants from 71 cross combinations were selected from F_3. Non-lodging and non-shattering were used as criteria in selecting 644 pedigrees from the F_5 planted at AVRDC. From 644 F_6 lines, 80 from 19 crosses outyielded local varieties SJ 1, SJ 2, and SJ 4. Out of the 80 lines, 17 F_8 lines from 9 crosses gave significantly higher yield than the local varieties during the rainy season. Among the seasons evaluated, the rainy season gave the highest seed yields at Praputtabat. Between different locations tested during the dry season ChiengMai gave the highest seed yield, but it was only next to Praputtabat. The year x genotype interaction variance for the dry season was highly significant for seed yield, indicating the necessity to evaluate the genotypes for more than one year. Similarly, significant variances for location, genotype and genotype x location for yield indicated the need to test in more than one location, and selection for location-specific genotypes may be effective. The mean yield of 20 varieties during three years in three locations was 1005 kg/ha. The three highest yielding selections gave a yield of 1218, 1158 and 1148 kg/ha compared to 990 kg/ha for SJ 1, 1130 kg/ha for SJ 2 and 1123 for SJ 4. Disruptive seasonal and locational selection may be useful in hastening the development of varieties adapted to specific seasons or locations.

GERM PLASM SHARING IN SOYBEAN IMPROVEMENT

K. S. McWhirter and I. A. Rose, Department of Agricultural Genetics, University of Sydney, N.S.W. Australia and Department of Agriculture Research Station, Narrabri, N.S.W. Australia

A full analysis of a soybean germ plasm collection for lines useful for improving yield and quality attributes is a mammoth task. One approach to simplifying this problem is to devise procedures which enable germ plasm analysis to be a cooperative venture of soybean breeders. We suggest two procedures for cooperative germ plasm analysis in soybean, and we offer lines from our breeding projects as contributions towards implementing these procedures.

The procedure of recurrent selection of recombinant inbred lines outlined by Compton (1968) is adapted for germ plasm analysis. The procedure involves: 1) making 100 single crosses among n [= 100] parent lines; 2) deriving, by single seed descent, a single inbred line from each single cross; and 3) yield testing 100 recombinant inbred lines and selecting 20 lines for recombination in a recurrent selection plan.

Soybean breeders could cooperate by donating to a common pool of lines one (or a few) recombinant inbred line(s) from each cross studied in their individual programs. Breeders implementing this procedure could start at step (3), by choosing a set of lines for yield test from the common pool. Availability of a large pool of lines of known pedigree origin would enable multiple sets of *n* parent lines to be analyzed, would reduce the time and effort required to reach the selection stage for each set, and would avoid duplication of effort. We have lines from 45 soybean crosses, involving 40 parent lines, available for contribution to the proposed pool of recombinant lines.

A second procedure involves the prediction of the properties of recombinant inbred lines from "adaopted" X "exotic" soybean crosses. We use the Hanson and Weber (1961) method of quantitative genetic analysis to obtain the measurements necessary for prediction of the properties of recombinant inbreds from soybean crosses. Useful crosses for breeding purposes are those predicted to give lines of high yield, or lines with desired combinations of quantitative characters. This procedure represents an application of statistical genetics to the problem of germ plasm assay.

Soybean breeders could cooperate in extending this procedure by donating sets of lines for a Hanson and Weber (1961) analysis of variability in many other programs. Sharing of sets of lines would avoid duplication of effort and greatly increase the number of crosses available for analysis in an individual program. We have available for sharing sets of lines structured as 2 lines from each of 30 different F_2 plants, from each of 20 crosses.

SOYBEAN [*GLYCINE MAX* (L.) MERRILL] IDEOTYPES IN TWO AGRO-CLIMATIC CONDITIONS

N. Mehrotra and B. D. Chaudhary, Haryana Agricultural University, Hissar - 125 004, Haryana, India

The aim of raising 30 randomly selected soybean genotypes at Hissar (alluvial plains) and Kangra (submontane) was to estimate phenotypic (r_{ph}) and genotypic (r_g) correlation coefficients and path coefficients for yield and its contributing characters to formulate specifications for plant ideotype and then verify the validity of adopting a uniform breeding approach for ecologically different regions. When the correlation coefficient study suggested the location-wise specifications for desirable soybean plant type, path coefficient analysis (r_g level) substantiated the utility of tall early flowering and early maturing plants with less primary branches, large pod size carrying few bold seeds and high pod number/plant for Hissar. The desirability at Kangra was for short, late flowering and early maturing plants with more primary branches, less pods/plant and large pods having more relatively small seeds. These characters either substantiated the useful direct effects or obviated the undesirable effects. With the high heritability (H) estimates for the days to flowering and maturity indicating their utility during selection, the low coefficient of variability (GCV) for flowering and maturity, plant height, seeds/pod and seed size needed sifting of genotypically diverse soybean populations. Further, primary branches and pods/plant along with seed size differing considerably for GCV and H estimates at both places suggested their differential response during selection. These evidences substantiated the utility of formulating different plant ideotypes for ecologically different regions. On this basis, future lines of soybean improvement programs have been outlined.

BREEDING SOYBEANS ADAPTED TO DOUBLE CROPPING

M. E. Pyle and G. R. Buss, Agronomy Department, Virginia Polytechnic Institute and State University, Blacksburg, Virginia 24061

Double cropping soybeans has become a large-scale practice in Virginia. All of the varieties being used have been selected under full-season growing conditions which are not typical of a double crop environment. Previous research indicates that the higher yielding full-season lines are not always the best yielding ones when double cropped. The purpose of this study was to examine the feasibility of selecting high yielding lines that are superior in performance in double cropping to the standard varieties.

In 1977, a number of F_2 and F_4 segregating populations were planted both as a full-season and no-till double crop following oats at the Eastern Virginia Research Station in Warsaw, Virginia. Single plant selections were made from the same populations at both planting dates. Four F_4 populations and seven F_2 populations were selected for further evaluation in 1978 based on the amount of seed available. Both hill plots and short rows (.91 m) were used to evaluate the F_4 selections. F_2 selections were evaluated only with short rows. A split plot design was used with a specific cross as the main plot and the sub-plots being progeny rows from the full-season or double crop selections. All plots were planted in plowed soil on June 23, 1978. Data were collected on plant height and width 42 days after planting. An early frost killed most of the plots before they were mature so they were given a relative maturity score based on the pod color two days later. Mature plant height and height of the lowest pods were also taken. Yield per plot and weight per 100 seeds were also analyzed.

Analysis of the data from F_4 rows did not show a significant planting date effect for yield. The planting date x cross interaction was significant at the 5 percent level indicating that the crosses reacted differently at different planting dates. The double crop selections tended to yield more than the full-season ones in three of the four populations. In the remaining population, the selections from the early planting had the highest average yields. Hill plots showed a significant date effect at the 1 percent level but no date x cross interaction. The double crop selections from all four populations significantly out-yielded the full-season ones. The reason for the different results in hill versus row plots is uncertain, but the data do indicate that planting date of the segregating populations can influence the selection results. It also appears likely that an improved double-crop variety can be selected from late-planted segregating populations.

AN IDEOTYPE FOR SOYBEAN

T. P. Yadava, Haryana Agricultural University, Hissar, India

Seed yield in soybean, like other crops, is a complex character, which in turn is the phenotypic expression of all the morphological, physiological and agronomic characters. Its improvement is the main object which is being attempted by breeders, physiologists, agronomists, pathologists, entomologists, microbiologists and biochemists at their own level. An integrated approach is required to achieve this goal.

Soybean is an important oil as well as protein crop. Breeders have tried to improve only the quantitative or economic characters, seed yield. Negligible research has been conducted on qualitative characters like oil and protein content. No plant ideotype has been formulated considering qualitative as well as quantitative characters. Separate plant types have been reported for varying environments and cultural practices.

In the present investigation, plant ideotypes have been formulated for biochemical, agronomical, physiological development and morphological characters under a range of environmental stress conditions. Breeding methodology has been discussed for plant ideotypes which will be ideal under different environments.

BREEDING FOR RESISTANCE TO YELLOW MOSAIC AND RUST DISEASES OF SOYBEAN

B. B. Singh, Department of Plant Breeding, G. B. Pant University of Agriculture and Technology, Pantnagar, Dist. Nainital, India

Yellow mosaic is one of the most serious diseases of soybean in India, Bangladesh, Sri Lanka and Pakistan. Soybean rust is a major disease in the whole of southeast Asia. Therefore, the major emphasis in soybean breeding programs at Pantnagar has been on the development of resistant varieties to these diseases. Through systematic screening of the world soybean germplasm, two lines resistant to yellow mosaic, PI 171443 and *Glycine formosana* (a variant of *G. soja*) and nineteen lines resistant to moderately resistant to rust were identified. Genetic studies have revealed that resistance to yellow mosaic is controlled by two pairs of recessive genes and the resistance to rust by one dominant gene. *G. formosana* appears to have a single dominant gene for resistance to yellow mosaic. The advanced breeding lines derived from segregating populations involving these resistant sources have exhibited high yield potential and resistance to yellow mosaic, rust and other foliar diseases. The details of these will be presented in the paper.

INDUCED GENETIC VARIABILITY FOR QUALITATIVE AND QUANTITATIVE CHARACTERS IN SOYBEAN

B. B. Singh and H. D. Upadhyaya, Department of Plant Breeding, G. B. Pant University of Agriculture and Technology, Pantnagar, Dist. Nainital, India

Ten varieties viz Bragg, Ankur Semmes, Hood, Lee, PK-71-39, UPSM-534, UPSM-229, Type-49 and UPSL-18 were each irradiated with 10 kr, 15 kr and 20 kr of gamma rays and the M_1, M_2 and M_3 generations were screened and evaluated for frequency, spectrum and inheritance of qualitative mutants. The magnitude of induced genetic variability for quantitative characters were studied in the M_3 generation. The induced qualitative mutants were of three types: 1) those showing chlorophyll deficiency, 2) those with leaf deformities, and 3) those showing sterility. Leaf mutants were more predominant with a mean frequency of

1.08 percent followed by sterile mutants (0.62 percent) and chlorophyll mutants (0.46 percent) with considerable varietal differences. UPSM-229 showed the maximum mutation frequency (5.26 percent), whereas T-49 had the lowest (0.77 percent). All mutants exhibited monogenic inheritance and the mutant character in each case was recessive.

The variances within irradiated populations were significantly higher than the variances of respective control populations for most of the quantitative characters indicating induced genetic variability. The magnitude of differences in variances for different characters varied from variety to variety. The estimates of heritability (broad sense) ranged from 0 to 51 percent for days to flowering, 0 to 58 percent for days to maturity, 0 to 80 percent for plant height, 0 to 49 percent for primary branches, 0 to 81 percent for pods per plant, 0 to 92 percent for seed per pod, and 0 to 80 percent for yield per plant. However, the mean performance of irradiated and control populations for all characters was not significantly different indicating that the mutations were random both in positive as well as negative directions.

A STUDY OF THE GENETIC CORRELATION BETWEEN LEAF INDEX AND NUMBER OF SEEDS PER POD IN SOYBEANS

H. S. Lee, Seoul National University, Suweon, Korea

The heredity of leaflet shape which was determined by the leaf index and the number of seeds per pod was studied in three cultivars and their F_1 and F_2 of soybean, *Glycine max.* (L.) Merr. The F_2 populations of crosses between Suweon 51, round leaflet, and SRF-300, long narrow leaflet, segregated in a 3 (round) to 1 (long narrow) ratio, indicating that the long narrow leaflet was controlled by a single recessive gene. The segregation of the number of seeds per pod in F_2 populations of crosses between Suweon 51 x SRF-300 showed a normal distribution. In this F_2 population, there was a highly significant correlation between leaf index and number of seeds per pod (r = 0.694).

HERITABILITY AND RESPONSE TO SELECTION OF PHYSIOLOGICAL TRAITS RELATED TO YIELD

R. Ecochard and M. H. Paul, E.N.S.A.T. University of Toulouse, 31 Toulouse, France

Breeding soybean genotypes adapted to suboptimal climatic conditions rests mostly on physiological criteria such as plant development and assimilation area, highly connected to yield (Planchon, 1978; Ecochard et al., 1978). Heredity of a few traits relevant on this point of view was studied in Toulouse in two independent non-selected hybrids segregating in F_3 and F_4. Both generations were grown as spaced plants, related by single seed descent. In addition, F_4 material was also analyzed on a plant mean basis in two replicated rows at normal density.

The scored characters were: development phase durations (emergence/pod formation/maturity), leaf area, yield and its components. Observed variances in F_3 and F_4, and covariances F_3 F_4 were calculated, and partitioned into additive, dominance, and environmental components. Heritability values could thus be estimated, whence the expected response, at the end of the selection cycle, to 10 percent selection intensity for each trait. We compared values obtained in the two families and whether the material would be spaced or densely planted.

The main results were: For both families as isolated plants, heritability values were consistently high (.4 to .8) for leaf area and earliness. They were low (below .1) for second development phase duration and weight per seed. The two families differed for heritability of yield, pods per plant and first phase duration. When the F_4 was grown under plant competition, heritabilities did not differ from the previous situation as a whole. Exceptions concerned the yield and especially pods per plant, the heritability of which was greatly reduced. The expected genetic progress resulting, at normal density, from a selection performed among spaced plants, showed more spread differences between characters, with total similarity among both families. The more responsive trait was leaf area (25 percent response), followed by yield (10 percent).

In conclusion, although the two above characters are positively correlated, genetically as well as environmentally, it is possible to associate in a breeding program an improved yield with a reduced foliage surface. This can be done by selecting out individual plants which, for a given maturity group, here end group I, show an early pod formation.

DISRUPTIVE SEASON AND LOCATIONAL SELECTION IN SOYBEAN FOR TAIWAN

A. T. Hung and S. Shanmugasundaram, Kaohsiung District Agriculture and Improvement Station, Pingtung, Taiwan and Asian Vegetable Research and Development Center, Shanhua, Tainan 741, Taiwan, Republic of China

Soybean, *Glycine max* (L.) Merr. is grown in Taiwan during the spring (Feb. to May) in Hualien and Pingtung; during the summer (June to Sept.) in Chiyai and Yunlin; and during the autumn (Sept. to Dec.) in Pingtung and Kaohsiung. None of the present varieties have sufficiently wide adaptability to be grown in all locations or seasons. Therefore, the Kaosiung District Agricultural Improvement Station, DAIS, in Pingtung is attempting to develop high yielding, early maturing, widely adapted varieties.

In the spring of 1978, 45 soybean lines resulting from 21 crosses were planted in randomized complete blocks, with two replicates. In Pingtung, 19 of the breeding lines outyielded the local check, Shih Shih variety widely used on farmer's fields. However, at Tainan only three of the lines outyielded the local check for that region, Tainung No. 4. The variances for genotype, location and genotype x location were highly significant for days to flowering, maturity, pods per plant, and 100-seed weight. The seed yield variances for location and genotype x location were highly significant, but genotypic variance was not significant. It is concluded that selection for yield should be done at specific locations. Oil and protein content were not significantly affected by location.

The seed yield was highly correlated with days to maturity, 100-seed weight and seed weight per plant at both locations. The 100-seed weight was negatively correlated to the severity of the soybean rust disease, *Phakopsora pachyrhizi*, Sydow. Varieties tolerant to soybean rust had larger seed size and higher seed yield than those which were susceptible to the disease.

IMPORTANT AGRONOMIC CHARACTERISTICS OF SOYBEAN CULTIVARS IDENTIFIED IN WORLD-WIDE ISVEX TRIALS

W. H. Judy, INTSOY, Department of Agronomy, University of Illinois, Urbana, Illinois 61801

The objective of the INTSOY soybean variety development program is to develop cultivars with high stable yield for tropical areas of the world. The INTSOY soybean breeding program is located in Puerto Rico. Cultivars from INTSOY and other breeding programs are first evaluated in the SIEVE trial at three sites. Selected cultivars are then evaluated at 15 to 18 sites in SPOT. The best cultivars are selected for ISVEX which is conducted by cooperators at more than 200 sites annually. The objective of ISVEX is to evaluate cultivars for wide environmental adaptability and to provide countries with improved cultivars for introduction or for breeding programs.

ISVEX contains 16 cultivars planted in four replications. Entries are separated into three groups— early, medium, and late maturing— with later genetic material planted nearest the equator. Cooperators return data on yield, days to flower and to maturity, nodulation, plant height, lodging, shattering, population, and seed characteristics. INTSOY provides oil, protein and statistical analysis of data. The responses to environment of the six highest yielding cultivars in more than 376 trials during 1973 to 1976 were tabulated. Three cultivars, Williams–III, Davis–VI, and Improved Pelican–VIII, were studied separately for the same sites and years. Sites were combined according to latitude (day length) and altitude (day- and night-time temperatures). Trends among the highest six and the selected three cultivars were similar.

For the three selected cultivars evaluated at decreasing latitudes, there was an increase in yield from 1679 to 2111 kg/ha and 1000 seed weight from 13.9 to 18.0 g; a decrease in plant height from 53 to 42 cm, in days to flower from 40 to 31, and in days to maturity from 104 to 93; and no change in percent protein in grain (~41.0 percent) and days from flowering to maturity (~64). With decreasing altitude at the same latitude, there was an increase in yield from 1323 to 2153 kg/ha. Days to flower decreased from 52 to 33, days to maturity from 123 to 95, and days from flower to maturity from 71 to 62. The response in plant height, 100 seed weight, and protein content was mixed. Analysis of the response of the individual cultivars indicated similar trends which different in magnitude among entires. Cultivars substituted by cooperators exceeded the yield of supplied entries in about 5 percent of the trials.

STUDIES ON THE BEHAVIOR OF TRUE-BRED MUTANTS OF SOYBEANS [*GLYCINE MAX* (L.) MERR.] IN NORTH ALGERIA

I. Nicholae and G. Ougouag, Institut National Agronomique, Department of Agronomy, El-Harrach (Alger), Algeria

The research in the production of soybean [*Glycine max* (L.) Merr.] has commenced quite recently in Algeria, in inducing artificial mutation by selecting favorable mutants from the collection of true-breds issued from seed irradiated in Romania. The biological materials utilized are 45 true-bred mutants of soybean. Eleven were retained after the first two years of experimentation for their resistance to environmental factors and for their satisfactory yield. These true-breds were mutants obtained from Romania after re-irradiation of the mutant true-bred B 107/10 (1972), issued from the American variety *Chippewa* (subjected to gamma rays-dose 30 Krads). The aim of the trials was to compare the mutant true-breds which have issued B 107/10, as well as other American and Chinese varieties. Experiments with different true-breds and varieties of soybean have been treated under dry (non-irrigated) and water channel conditions existing at Mitidja and Haut-Cheliff, respectively. Soybean plots (6 by 20 M^2) were planted manually in April in a randomized block design. Experiments were replicated three times. The authors analyzed the different soybean mutants concerning their principle morphological (young seedlings, adult plants, pods and the grains), physiological (resistance to spring frost, dryness, lodging, and maturity) and biochemical characters (the production of proteins and fats).

The morphological studies of the soybean true-breds have permitted the identification of the mutant characters of the different vegetative organs affected by the irradiation with regard to the control (the initial variety non-irradiated) and to establish detailed descriptions of new soybean forms. Certain mutant morphological characters, especially absence of petioles (sessile leaves), short angles between petiole and the stem, the size of leaves, etc., presented a particular interest in increasing the density of the plants or for the high yield of mass vegetation. It is worthwhile to note that most of the true-bred mutants are different by several characters in comparison to the initial variety. Generally the true-breds and the varieties investigated were sensitive to soil temperatures below 8 C and did not germinate. Only two true-bred mutants exhibited a resistance to the freezing cold (B 62/15 and B 10/37) and were capable of germinating easily at a temperature of 6 to 8 C. The examination of the drought character is very important for the soybean extension in Algeria. The B 62/15 and B 10/37 were very resistant to drought. Certain true-breds were more resistant to lodging than others. The vegetative cycle of the true-bred required less time for maturity (20-30 days) than that of the initial variety, especially B 79/37, B 79/521, and B 10/610. The amount of protein examined in the dry matter of the true-breds varied from 35 to 45 percent. Only the true-bred (B 10/37) was superior to the control (the rate of protein production was 45 percent in relation to the control, which registered 44 percent). The amount of oil varied from 14 to 23 percent. The whole lot of the true-breds were superior to the controls. At the end of four years of experimentation, the true-breds which gave interesting yields are B62/15, B 10/55, B 10/37 and B 10/19. It is our opinion that the best true-breds were mutants which attained maturity before the hot summer and the Sirocco.

Mutants were selected which were adapted to the climatic conditions in Algeria. The true-bred, B 62/15, has been officially recognized under the name Cerag No. 1.

PHYSIOLOGY

Invited Papers

AMINO ACID LOADING AND TRANSPORT IN PHLOEM

L. E. Schrader, University of Wisconsin, Madison, Wisconsin 53706

The mechanisms controlling loading and transport of amino acids and sucrose were compared in soybeans [*Glycine max* (L.) Merr. cv. Wells]. After steady-state labeling with $^{14}CO_2$ for two hours, about 70 percent of the [^{14}C] recovered in the fed source leaf was in sucrose and about 15 percent was recovered in amino acids. In the transport pathway (petiole of fed leaf and stem), about 90 percent of the recovered [^{14}C] was in sucrose and 5 percent was in amino acids. Most of the [^{14}C] in amino acids from the transport pathway was in only five amino acids. These findings suggest that some selectivity exists in the vein loading of both sucrose and amino acids.

Heat girdling the petiole of the source leaf to disrupt only the phloem stopped translocation of [^{14}C] assimilates from the source leaf. This indicates that both sugars and amino acids are transported from the source leaf via the phloem.

Radioactive amino acids and/or sugars were exogenously applied to abraded areas of soybean source leaves, and transport to a younger sink leaf was monitored. When glucose, fructose, and sucrose were applied to the abraded spot, sucrose was specifically loaded. Leucine, lysine, glutamate, serine, and γ-amino butyrate were translocated with similar velocities and mass transfer rates on a molar basis. Amino acid vein loading showed little selectivity or specificity on the basis of charge, position of the amino group, or natural occurrence in the phloem.

The loading of sucrose and leucine was concentration-dependent. Leucine loading showed a tri-phasic saturation response whereas sucrose showed a bi-phasic saturation response. For leucine, the first phase was from 0 to 4 mM (K_m = 3 mM), the second from 4 to 40 (K_m = 21 mM), and the third from 40 to 100 mM (K_m = 52 mM). For sucrose, the first phase extended from 0 to 110 mM (K_m = 35 mM) and the second phase was from 110 to 400 mM (K_m = 169 mM). The mass transfer rate of leucine translocation was not affected by exogenous sucrose at concentrations as much as 10 times higher than that of leucine. This lack of competition for carrier sites, the presence of many saturated phases in the concentration response curves, and the differing kinetic properties suggest that transport of sugars and amino acids is mediated by separate carriers.

Both leucine and sucrose translocation were sensitive to treatments that have been reported to inhibit sucrose vein loading. All phases of the concentration response to leucine and sucrose translocation were sensitive to inhibition by 5 mM dinitrophenol. Other effective inhibitors of sucrose and amino acid transport were KCl, PCMBS, and CCCP. These inhibitor studies suggest that amino acid loading is an active process.

NITROGEN TRANSLOCATION IN THE XYLEM OF SOYBEANS

D. W. Israel and P. R. McClure, USDA, SEA and Soil Science Department, North Carolina State University, Raleigh, North Carolina 27650

Analysis of sap bleeding from cut stems has been used to study the transport of nitrogen in the xylem of soybean (*Glycine max*) plants grown in both the greenhouse and the field. Predominate forms of nitrogen transported in xylem sap vary depending on the source of nitrogen supplied to the plants. During vegetative through mid pod fill growth stages, ureides (allantoin, allantoic acid, and urea) contained an average of 78 percent of the total N in sap collected from plants dependent on N_2 fixation, while amino acids contained about 20 percent. During the same period of growth, ureides contained an average of 6 percent of the total N in sap of non-nodulated plants dependent on assimilation of nitrate, while amino acids and nitrate contained 36 and 58 percent, respectively. Under all growth conditions examined, asparagine contained 50 to 70 percent of the nitrogen in the amino acid fraction.

When nodulated plants were supplied with increasing levels of nitrate, sap ureide content declined in proportion to the degree of inhibition of nodulation and of total plant N_2 fixation (C_2H_2 reduction) activity resulting in a positive correlation ($r = 0.989$) between percent of total nitrogen input by N_2 fixation and percent total sap N as ureides. Inhibition of N_2 fixation by continuous exposure of nodulated root systems to saturating levels of acetylene (0.1 atm) decreased the sap ureide concentration by about 88 percent within 24 hr. This suggests that synthesis and transport of ureide is closely coupled to current N_2 fixation and that turnover of ureides in nodule transport pools is relatively rapid. The correlation between sap ureide composition and N_2 fixation activity and the close coupling of ureide transport to current N_2 fixation suggest that it may be feasible to use sap ureide content as an index of relative contributions that N_2-fixation and uptake of soil and fertilizer nitrogen make to nitrogen input at specific growth stages.

Current physiological and biochemical data suggest that most of the ureide-N transported in the xylem is synthesized in the nodule and that ureide-N is metabolized to support plant growth and development. The domination of ureides in transport of nitrogen in the xylem suggests that assimilation of nitrogen in soybean nodules after fixation from N_2 is more complex than originally thought. Determination of the energy requirement per unit nitrogen transported as ureides must await elucidation of the details of their biosynthetic pathway.

COUPLING BETWEEN PHOTOSYNTHESIS AND N_2 FIXATION

J. G. Streeter, H. J. Mederski and R. A. Ahmad, Ohio Agricultural Research and Development Center, Wooster, Ohio 44691

Several experimental approaches (source-sink manipulations, CO_2 enrichment, alteration of the light environment) have established a close coupling between photosynthesis (PS) and nitrogen fixation (NF) in soybeans. Although carbohydrates constitute the obvious linkage between PS and NF, little information is available on the distribution and metabolism of carbohydrates in nodules.

There is often no apparent relationship between total non-structural carbohydrate content of nodules and NF. This may be due to the fact that much of the non-structural carbohydrate in nodules is in the form of starch and cyclitols. These components do not appear to be in the "mainstream" of carbohydrate metabolism in nodules, but are utilized, presumably during periods when the influx of sucrose is limited. In studies where NF has been altered by various treatments (variable light intensity, temperature, water stress), only sucrose and D(+)-pinitol content of nodules is consistently correlated with NF. The physiological basis (if any) for the relationship between pinitol and NF remains unknown.

The quantity of carbohydrate consumed in legume nodules per gram of nitrogen fixed has been determined experimentally. The theoretical energy requirement for NF can be calculated from *in vitro* measurements of NADPH and ATP required by nitrogenase. Comparison of theoretical and actual energy requirements indicates that perhaps a 2- or 3-fold increase in the efficiency of NF (g N fixed/g glucose) might be achieved. The meager amount of evidence which has a bearing on the feasibility of achieving this goal will be reviewed. It is concluded that much more information on carbohydrate metabolism will be needed before a final judgement can be made.

REGULATION OF SENESCENCE

L. D. Noodén, Botany Department, University of Michigan, Ann Arbor, Michigan 48109

During pod fill, soybean plants undergo rapid degeneration (monocarpic senescence) causing a loss of productive capacity and eventually death. The most obvious manifestation of these changes is yellowing of the leaves, and this is our main measure of senescence. This degeneration of the leaves is probably also a key aspect of monocarpic senescence.

Although vegetative growth ceases early in the reproductive phase, this seems to be separate from the senescence process. Neither age nor size are determinants of longevity in soybean, for prevention of flower development (with a noninductive photoperiod) allows the plants to live far longer and grow far beyond normal size. Since deseeding prevents monocarpic senescence, the developing seeds control the senescence process; we will refer to this influence of the seeds as the senescence signal without making a commitment to any particular mechanism. The senescence signal shows limited mobility (primarily downward if at all) and is produced toward the end of seed growth when nutrient accumulation is probably complete. Any of a variety of surgical manipulations will allow pod development to proceed without senescence. For example, plants may be defoliated and depodded leaving only one leaf and one pod cluster separated by three internodes. When the leaf is above the pod cluster, the leaf does not senesce; however, when it is below the pod cluster, it senesces in about half the time. The seed yield is the same whether the pod cluster is above or below the leaf, but the yield is greater than an equivalent pod cluster on an unmodified plant. Thus seed development can occur without causing senescence; those nutrients (or hormones) which the seeds require from the rest of the plant can be obtained without killing the plant. Although nutrient (or hormone) diversion or drain from the vegetative parts to the developing pods does occur during monocarpic senescence, these do not seem to be primary causes of senescence, i.e. the senescence signal. I favor the idea that the seeds produce one or more hormones which induce foliar senescence.

Rather little information has been published on hormonal control of monocarpic senescence in soybeans. Analyses based on bioassays and gas-liquid chromatography of partially-purified extracts show that foliar abscisic acid-like activity rises to a high level during the period when the seeds are inducing senescence, and cytokinin-like activity drops prior to this time. Foliar applications of abscisic acid accelerate monocarpic senescence but do not induce senescence in podded plants. Both auxin (naphthalene acetic acid) and cytokinin (benzyladenine) applications can retard monocarpic senescence; however, together they can actually prevent it without interfering with pod development.

EFFECT OF POD FILLING ON LEAF PHOTOSYNTHESIS IN SOYBEAN

W. A. Brun and T. L. Setter, Department of Agronomy and Plant Genetics, University of Minnesota, St. Paul, Minnesota 55108

Soybean yield may be considered as a product of the photosynthetic rate, integrated over time, and the partitioning of the resulting photosynthate between physiological and morphological yield components. Published reports indicate strong interactions between activities in photosynthetic sources and sinks. The observed interactions between the pod filling process and the photosynthetic activity in the leaves has been interpreted to be caused either by nutrient concentration gradients between the pods and the leaves, or to be due to hormonal signals between the pods and the leaves. All of the major endogenous hormones have been implicated as possible regulators of this interaction.

In our laboratory we have observed that soybean pod removal causes partial inhibition of leaf photosynthesis. This inhibition is caused entirely by stomatal closure and is not related to photosynthetic ability in the mesophyll. We hypothesize that the filling pod acts as a sink for stress produced abscisic acid (ABA) in the leaves. Thus when the pod is removed, ABA accumulates in the leaf causing stomatal closure. We have previously demonstrated that developing soybean seed contain large amounts of ABA. We now demonstrate that removing the pods or interrupting the phloem tissue in the petiole causes a marked increase in the free ABA content of the leaf blade and marked stomatal closure.

POSTPONEMENT OF SEVERE WATER STRESS IN SOYBEANS BY ROOTING MODIFICATIONS: A PROGRESS REPORT

H. M. Taylor, SEA/AR, USDA and Agronomy Department, Iowa State University, Ames, Iowa

Distribution of vegetables over the earth's surface is controlled more by water availability than by any other single factor. Consequently, a tremendous amount of research is conducted on methods to increase the quantity and effectiveness of the total water supplies. The objective of this paper is to present a progress report of efforts to increase the total water supply available to soybeans by modifying their root systems. The paper describes briefly a conceptual model of water uptake by a plant root system. Components of the model are transpiration rate, volume of rooted soil, root length density, water uptake rate per unit root length, plant water potential at crown level and axial resistance within the xylem elements.

When experimental results are incorporated into it, the model indicates that increases in rooting depth are the most likely root-modification way to increase total water supply for soybeans grown in Iowa. Experiments show that rooting depth can be increased by: 1) selecting plants with more rapid taproot elongation rates during vegetative stages, 2) selecting plants that will allow rapid root elongation during pod-filling, 3) fertilizing deficient soil, and 4) removing a water table. Experiments to evaluate each of these possibilities are discussed in the paper.

BIOASSAYS TO DETECT CHEMICAL EFFECTS ON FLOWER ABORTION IN SOYBEAN

A. Huff and C. D. Dybing, Plant Science Department, University of Arizona, Tucson, Arizona, and SEA/AR, USDA, South Dakota State University, Brookings, South Dakota

To test the hypothesis that plant hormones or other chemicals are involved in flower shedding, bioassays are needed which accurately mimic *in vivo* abortion. The present work was concerned with developing and testing abortion bioassays for soybean. The principal bioassay tested was an *in situ* assay depending on chemical replacement of the abortion-increasing effect of lower flowers on those flowers borne on upper parts of the raceme. Plants of 'Clark,' isoline E_1t, were grown in growth chambers in 13-hr photoperiods, 29/24 C (day/night) temperature cycles, and 450 $\mu Em^{-2}sec^{-1}$ (400-700 nm) irradiation levels. The growth medium was peat:vermiculite (1:3, v:v) watered daily with nutrient solution. The three lowermost flowers on each raceme at the 7th, 8th and 9th nodes were excised at the base of the calyx on the day of the first flower's anthesis. Pedicels for these three flowers were removed the following day by brushing with a small glass rod. Lanolin containing test chemicals were then applied to pedicel scars and fourth flowers were monitored daily to detect abscission or pod set. Abscission was defined as the day the fourth flower was missing or would fall of it touched lightly. The protrusion of the ovary beyond the calyx was taken as the first clear indication of pod set. Known hormones, some possible organic nutrients, and extracts prepared from developing pods were tested by this system at rates of 0.01, 0.1, 1.0, and 2.0 percent in lanolin.

IAA was the only hormone tested by the *in situ* assay that promoted flower abscission. IAA also caused bending of racemes and suppressed petal opening. Extracts from young pods gave variable results; two out of three were as effective as IAA in inducing abscission but caused no growth abnormalities. Developing pods left intact at the three lower positions on the raceme also stimulated abscission of the fourth flower without growth abnormalities. The GA at 0.01 and 0.1 percent increased pod set. Benzyladenine, glycine, sucrose, abscisic acid, and indoleacetaldehyde had no effect in the bioassay. *In vitro* culture techniques were also tested for utility as abortion bioassays. Culture of developing ovaries appeared more promising for such assays than culture of entire racemes.

Contributed Papers

HERITABILITY OF CANOPY PHOTOSYNTHETIC CAPACITY AND ITS RELATIONSHIP TO SEED YIELD

S. A. Harrison, H. R. Boerma, and D. A. Ashley, Agronomy Department, University of Georgia, Athens, Georgia 30602

Previous research has shown variation among and within maturity group V, VI, VII and VIII cultivars for canopy apparent photosynthesis (CAP) during the podfill period in soybeans, *Glycine max* (L.) Merrill. This work also indicated a positive relationship between point measurements of CAP during podfill and seed yield. The objectives of this study were to: 1) determine the heritability of CAP, and 2) determine the relationship between CAP and seed yield in progeny from a cross between parents differing in their CAP rates.

Thirty-four F_3 derived lines each from two crosses ('Dare' x 'Forrest' and 'Tracy' x 'Davis') were grown in replicated field plots in 1977 and 1978. CAP measurements were obtained during the podfill period with a 1 m Lx1 m Wx1.6 m H, Mylar covered, portable chamber with a removable lid. CAP was determined from the decrease in CO_2 concentration in the chamber over time. Each plot consisted of three rows 2.1 m long spaced 91 cm apart. CAP was measured on the center row. In both years seed yield was measured separately on the center row and the border rows. In 1978 three additional replicated experiments were grown to provide a better estimate of seed yield.

Heritability estimates for CAP over years from both crosses ranged from 36 to 66 percent by the variance component and standard unit methods based on the mean of two measurements each year during the podfill period. In 1977 the correlation coefficients between CAP and seed yield were 0.50** and 0.48** for the Dare x Forrest and Tracy x Davis crosses, respectively. The relationship between CAP and seed yield in 1978 and the use of CAP for early generation selection of heterogeneous lines will be discussed.

DRY MATTER DISTRIBUTION DURING THE ONTOGENY OF 13 SOYBEAN GENOTYPES

G. R. Buss and L. H. Aung, Virginia Polytechnic Institute and State University, Blacksburg, Virginia 24061

Interest has been shown in breeding crops for more efficient conversion of photosynthate into harvestable plant parts. In order to accomplish this, the plant breeder must have sources of genetic variation for the trait in question. The objective of this experiment was to obtain an estimate of the genetic variation present in the relative growth rate (RGR) and net assimilation rate (NAR) of a group of soybean cultivars and lines. The partitioning of dry matter to the various plant organs was also investigated.

Thirteen cultivars and lines were grown in 1978 at Blacksburg, Virginia. They ranged in maturity from Group III to Group V and were planted in a randomized complete block design with six replications. A plot consisted of three 6.1 meter rows spaced .91 meters apart. Samples of six plants were taken from each plot on four sampling dates at roughly three-week intervals through the growing season. Dry weights were obtained for stems, leaves, petioles, and pods for both the main stem and branches. Seed yields were determined by harvesting the center row at maturity. There appeared to be very few significant differences in RGR among the cultivars evaluated. More significant differences between cultivars were observed for NAR, but the differences were not large. Neither RGR or NAR appear to be closely associated with mature seed yields. These preliminary results suggest that genetic variation in RGR and NAR in soybeans is not large in comparison to the environmental variation. Thus the plant breeder would probably find progress from selection for these traits to be very slow.

EFFECTS OF TEMPERATURE ON PHOTOSYNTHESIS IN SOYBEAN: GENETIC VARIABILITY AND PROSPECTS FOR NEW VARIETIES FITTED TO HIGH AMPLITUDE TEMPERATURE CHANGES

C. Planchon, Ecole Nationale Superieure Agronomique, 31076 Toulouse Cedex, France

The search for a better adaption of soybean to the unstable climatic conditions of some growing areas has led to the investigation of the response of photosynthesis to temperature in various cultivars, which was analyzed as follows: 1) effect of growth temperature on the gas exchange rates of leaves, 2) effect of short-term exposures to high or low temperatures on leaf photosynthetic activity, and 3) effect of long-term exposures to high or low temperatures (which are markedly different from the growth temperature) on photosynthesis in leaves whose growth is completed. The investigation was carried out on a series of cultivars originating both from America and Europe. Net photosynthesis was determined by I.R. analysis.

Marked differences in response to photosynthesis were observed for the various soybean cultivars grown at two different temperatures: 22 and 30 C. The American varieties had a better photosynthetic activity when the growth temperature was 30 C, whereas the European cultivars were more efficient at 22 C. The temperature which corresponds to the maximum photosynthetic rate is closely related to the growth temperature for most varieties, except for the Altona and GSZ 3 cultivars which show a poor adaptation. The effect of short-term exposures to temperatures differing from the growth temperature was a marked decrease in photosynthetic rates at low temperatures, mainly for the European cultivars. When the temperature was 10 C lower than the optimum temperature for photosynthesis, the decrease was about 30 percent.

The effect of long-term exposures to temperatures differing from the growth temperature showed that: 1) Soybean photosynthesis is markedly affected by low temperatures of long duration, which act on leaves whose growth had occurred under a higher temperature regime. This point is important in areas where warm periods corresponding to leaf growth are followed by cooler periods corresponding to the seed filling. 2) However, the structures developed during leaf growth under lower temperatures are at least as efficient as the ones built under higher temperatures. A wide genetical variability was observed in the whole set of responses of photosynthesis to temperature in soybean. The search for types that are less sensitive to the long-term effects of low temperatures during the filling period should lead to new varieties better fitted to unstable climatic conditions.

CORRELATION BETWEEN PHOTOSYNTHETIC RATE AND SPECIFIC LEAF FRESH WEIGHT, RUDP CARBOXYLASE ACTIVITY IN SOYBEAN VARIETIES

I. Watanabe, Tohoku National Agricultural Experiment Station 019-21, Kariwano, Akita, Japan

Specific leaf fresh weight (FW/dm^2), which is highly correlative with leaf thickness, has been reported to be correlative or noncorrelative with photosynthetic rate (P/dm^2) depending upon the materials used. The experiments were conducted in order to explain coherently these apparently contradictory results. The P/dm^2 was divided into two components, quantity factor and quality factor; $P/dm^2 = FW/dm^2 \times P/FW$, where P/FW is photosynthetic rate per unit fresh weight of leaves. This term itself cannot be measured but only calculated. Hence, RuDP carboxylase activity per unit fresh weight of leaves (RuDP/FW) was measured as an indication of P/FW. The P/dm^2 was measured by Toshiba-Beckman Infrared Gas Analyzer under saturating light of 50 klx. In each experiment, 15 or 16 varieties were used and 4 or 5 leaves from different plants of each variety were measured for P/dm^2 and FW/dm^2. The RuDP activity of the leaves used for photosynthesis was measured using crude enzyme extract from a portion of the leaves. Incorporation of ^{14}C into organic substances by the carboxylase, with substrates RuDP and labeled $NaH^{14}CO_3$, was measured by low back gas flow GM counter.

Upper leaves had the lowest FW/dm^2 and the highest CV of FW/dm^2 among varieties. With these leaves, correlation coefficients between P/dm^2 and FW/dm^2 were the highest and significant. They were larger than those between P/dm^2 and P/FW. Primary leaves, in contrast, had the highest FW/dm^2 and the lowest CV of FW/dm^2. With these leaves, correlation coefficients between P/dm^2 and FW/dm^2 were the lowest and nonsignificant. They were far smaller than those between P/dm^2 and P/FW. The correlation coefficient between P/dm^2 and RuDP/FW was not significant in upper leaves. It increased as leaf position lowered and became significant in primary leaves. These results suggest that the quantity of leaf mesophyll is the more important factor for the difference of varietal photosynthesis in upper leaves and that the quality of leaf mesophyll is the more important factor in lower leaves.

RELATIONSHIP OF SOYBEAN YIELD TO SEED FILLING PERIOD

K. J. Boote, T. R. Zweifel, A. M. Akhanda, and K. Hinson, Agronomy Department, University of Florida, Gainesville, Florida 32611

Genetic variability for length of seed filling period offers potential for increasing soybean yield, provided breeders can screen easily for longer filling period within the constraints of length of season and along side of other selection criteria. Other researchers have shown that only a small amount of carbon fixed prior to seed fill ends up in soybean seed; thereby indicating the importance of carbon assimilation during seed fill. Selecting for higher canopy photosynthetic rates during seed fill is desirable but difficult. A better approach may be to select for longer seed filling periods and thereby increase the total of carbon assimilated during seed fill. An important factor affecting the filling period is the daily seed N requirement as related to N remobilization from leaves and subsequent photosynthetic decline.

Our objective was to evaluate the relationship of seed yield to filling period for three different sets of soybean genotypes grown in very different environments at Gainesville, Florida. Set one was 13 genotypes in maturity groups (MG) 00 to IV planted in 30-cm rows in March 1976 and harvested in June-July. Set two was 8 full-season genotypes in MG VI, VII, VIII planted 16 June 1977 in 90-cm rows. Set three was 84 soybean genotypes in MGVIII, IX, X planted 15 July and 3 August 1977 in 41-cm rows. Good cultural and irrigation practices were followed. The length of seed filling period was visually estimated as days from first bean swell (R5) to first yellowing pods (near R7) or to 95 percent mature pods (R8).

Seed yield was positively correlated with length of filling period in all three studies. Yield of spring-planted 1976 soybean cultivars increased at a decreasing rate as filling period (R5 to R8) increased from 32 to 52 days. The average rate of yield increase was about 50 kg/ha per day increase in filling period. Results with full season genotypes in 1977 indicated that low yield of obsolete cultivars and high protein cultivars was associated with filling periods (R5-R7) up to a week shorter than those of adapted varieties Bragg and Cobb. Over the range of yield response to filling period, yield was increased at least 100 kg/ha for each day increase in filling period. A linear regression of yield to filling period (R5-R7) for cultivars planted 15 July and 3 August 1977 showed that yield was increased 59 and 62 kg/ha, respectively, for each day increase in filling period. These results demonstrate a positive association of seed yield to filling period in a wide range of soybean genotypes planted in three different environments. While these results certainly indicate potential for future yield improvement by selecting for filling period, they also suggest that breeders have already developed cultivars with long filling periods, merely by selecting for yield.

CHANGES IN SEED COMPOSITION DURING POD FILLING

P. W. G. Sale, L. C. Campbell, and O. G. Carter, Department of Agronomy and Horticultural Science, University of Sydney, N.S.W. Australia 2006

While other investigators have studied the accumulation of organic components in developing soybean seed during podfilling, there is a lack of information on nutrient accumulation in the seed. The purpose of this study was to examine nutrient levels in developing soybean seed as well as the oil and dry matter content. Each week for the duration of the 12-week podfilling period, four samples of tagged pods were collected from a crop of soybeans, cv. Lee, that was grown at Camden, N.S.W., under conditions of adequate nutrition and moisture. The pods developed from flowers tagged on the 10th to the 12th of February, and were from the top three nodes within the canopy. The seed was analyzed for N, P, K, Ca, Mg, Fe, Mn, Zn, Cu and oil.

All changes in seed composition discussed below were significant ($p < .05$). Maximum nutrient content (mg) levels occured at the end of the 10th week, except for those of K and Ca which occurred at the end of the 11th week, when maximum seed dry weight occurred. Maximum oil content occurred at the end of the 9th week, then decreased during the 11th and 12th weeks. Nutrient concentrations (percent of dry weight) in the seed fell during the 4th week, except for N, Mg, and Fe, which fell during the 5th week, and for K which did not fall until the 6th week. Minimum concentrations occurred at the end of the 7th week for most nutrients, except for Mg and N which occurred during the 5th week, and for K and Ca which occurred at the end of the 9th week. Increases in nutrient concentrations occurred during the 8th to 10th week, except for Mg, which increased from the 6th to the 10th week, and for K and Ca which increased from the 10th to 11th week.

The general pattern was that nutrient concentrations fell to a minimum then increased towards the end of podfilling. The stage of seed development when the concentration minima were reached, and when the concentration increased, varied with the nutrient. Oil concentration increased rapidly from the 3rd to the 6th week, remained constant for three weeks and then fell during the 11th week. The N concentration fell during the stage of rapid increase in oil concentration, but increased during the period when oil concentration remained steady. This indicates that the oil:N accumulation ratio decreased as podfilling proceeded past the 6th week.

CARBOHYDRATE DISTRIBUTION IN SOYBEAN PLANT PARTS AT DIFFERENT MATURITY STAGES AND LIGHT INTENSITIES

B. M. Coggeshall, H. F. Hodges, and T. K. Porter, Mississippi State University, Mississippi State, Mississippi 39762

Field experiments were conducted on soybeans, *Glycine max* (L.) Merr. cv. Tracy. The purpose was to determine the carbohydrate distribution among several plant parts at different maturity stages, and to determine the effect of a 92 percent reduction in light intensity on the carbohydrate balance within the plant. This information was used to determine the amount of 'reserve carbohydrates' that are normally present in the soybean plant at different growth stages and the extent to which these "reserves" can be mobilized and used by the plant. The experiment was conducted during four maturity stages: V8, vegetative plant; R2, flowering plant; R5, beans beginning to develop inside the pods; and R6, pods contain full size green beans. The plant parts collected were: the top three fully expanded leaves; bottom three green leaves; apical meristems; pods; top, middle, and bottom stems; and roots. Two light intensities examined were full natural sunlight, and 92 percent reduction in sunlight intensity caused by plastic mesh shades. Ten plants were collected per experimental unit at 7:00 a.m. three days after shade initiation. The tissue was freeze dried, ground, and analyzed for soluble sugar and starch concentration by a phenol sulfuric acid assay.

The highest starch concentrations under normal light intensities were generally in the leaves, pods and apical meristems. However, upper and middle stem sections during early podfill, R5, contained high levels of starch, therefore becoming an important reservoir during this growth stage. Sugar concentration during the vegetative and flowering stages was highest in the upper stems. Thus, when the plant is undergoing vegetative growth much of the sugar is available in the upper stems for translocation to vegetative meristems. Pods had the highest sugar concentration (mg sugar/g dry wt) during R5 and R6 stages. The large number of pods distributed over the entire plant are very efficient in extracting sugar from the source leaves and stems. The greatest percentage of starch depleted by shading was in the vegetative tissues, especially in the top and bottom leaves. This indicates a larger amount of available carbohydrate in young tissues. As the plant matures from vegetative to reproduction stages more starch appears to be incorporated into tissues or is unavailable for mobilization. This inability to mobilize the starch may be due to the absence of hydrolytic enzymes. Pod starch concentration under shaded conditions averaged between 72 and 87 percent of the control pods in the light, while the apical meristems were between 45 and 55 percent of the control. Pods accumulate storage starch in seed amyloplasts while apical meristems temporarily accumulate starch for subsequent mobilization and growth. Bottom leaves and roots during R2 stage and the top stem and bottom leaves during R5 maturity stage become important temporary storage sites because shading reduced the concentration of starch to the greatest degree in these plant parts.

REVERSIBLE ENHANCEMENT OF RATES OF SOYBEAN PHOTOSYNTHESIS THROUGH SHADING OF SIDE LEAFLETS

M. M. Peet and P. J. Kramer, Department of Botany, Duke University, Durham, North Carolina 27706

The rate of photosynthesis of soybean leaves is reported to be influenced by the demand for assimilates. To investigate this we grew Fiskeby V, a small determinate soybean cultivar, under 12-hr photoperiods in 23/17 C growth chambers with 50 percent RH and 450 $\mu E{\cdot}m^{-2}{\cdot}sec^{-1}$ PPFD. The photosynthetically active leaf area was decreased about 70 percent by shading with aluminized mylar either all side leaflets or all side leaflets and all pods. After treatment imposition, photosynthesis and transpiration were monitored in the unshaded center leaflet of the 5th trifoliate leaf for 18 days with an infrared gas analyzer in the differential mode and a dew point hygrometer. The rate of photosynthesis of the unshaded leaflets increased 50 percent within two days after shading and remained higher than controls until the end of the experiment. This increase in rate compensated for one-half of the loss in net carbon assimilation which would have occurred if there had been no enhancement of photosynthesis of the unshaded leaflets. Shading of leaflets decreased the time required to reach maturity. Shading of pods as well as leaves delayed pod development and senescence of plants for 2 to 3 weeks, but did not reduce the seed yield compared to plants with leaves but not pods shaded.

In a subsequent experiment reversibility of the enhancement of photosynthesis in unshaded leaflets was demonstrated by reversing the treatments after one week. Shades were removed from the side leaflets on half of the shaded plants and applied to side leaflets on half of the control plants. Within two days, rates of photosynthesis had decreased to control rates in the uncovered plants and increased in the recently shaded plants to the same rates as those continuously shaded. Pod and seed weights were not significantly decreased by one week of shading, but shading after one week through maturity decreased weights 25 percent. Plants shaded continuously in this experiment had 36 percent less pod and seed weight than controls. These experiments demonstrated that the rate of photosynthesis of soybean increases significantly when the photosynthetically active leaf surface is decreased, but the increase does not completely compensate for the loss in photosynthetic surface. The enhancement was reversible after one week. Leaf shading increased the speed of maturation. Pod shading delayed maturity, but did not significantly affect the rate of photosynthesis or day matter produced.

PHOTOPERIODIC RESPONSE OF FLOWERING IN DECAPITATED SOYBEAN PLANTS

S. Shanmugasundaram, Wang Chao-chin, and T. S. Toung, Asian Vegetable Research and Development Center, Shanhua, Tainan 741, Taiwan, Republic of China

Research has demonstrated that photo-sensitive and photo-insensitive soybean, *Glycine max* (L.) Merrill, cultivars can be distinguished using the "decapitation technique." Use of the "decapitation technique" showed that photoperiodic response is inherited. The objective of the present investigation was to understand the possible translocation of flowering substance from one branch to the other and its influence on the time to flowering and further development of the two branches in a decapitated plant. We used decapitated, two-branched soybean plants from photo-sensitive Acc. G2120 and photo-insensitive Acc. G215 to study the response to 10- and 16-hour photoperiods. There were 15 treatments. The first two treatments used whole plants. Treatments 3 to 15 used decapitated two branched plants. In treatments 3 to 15 donor and receptor branches were either completely defoliated or a definite number of leaves were retained.

The photoperiodic responses of decapitated two-branched sensitive Acc. G2120 and insensitive Acc. G215 soybean cultivars subjected to differential photoperiods were different. The receptor branch of sensitive cultivars flowered when they were either completely defoliated or had less than four leaves. Indirect evidence indicated the interplay of both flower-promoting and flower-inhibiting processes in the sensitive cultivar. The receptor branch of the sensitive cultivar developed pods only when it was defoliated or had only one trifoliate leaf. The number of flowers and pods produced were quantitatively proportional to the number of trifoliate leaves. A definite number of leaves saturated this requirement and decided the ceiling. Pod development on the sensitive cultivar requires additional photoperiodic stimulus and photosynthetic source.

The insensitive Acc. G215 produces flowering-promotor regardless of the photoperiod. Therefore, the times to flowering and maturity of the two branches of G215 were not affected by the differential photoperiodic treatments. The receptor branch without leaves can receive a stimulus to produce only the minimum number of flowers, even though the donor had all the leaves. More flowers and pods are produced in the 16-hour than in the 10-hour photoperiod. The number of pods produced are proportional to the number of leaves present, whether on the donor or on the receptor branch. In conclusion, the differential photoperiodic response of the decapitated two-branched plants is similar to that of the whole plants.

VALIDATION OF A SOYBEAN YIELD MODEL, BASED ON PHENOLOGY AND CLIMATE

P. G. Jones and D. R. Laing, Department of Agronomy and Horticultural Science, University of Sydney, Sydney, Australia

There have been many studies of soybean phenology, of yield components and of the effects of climate upon yield, but few so far on the combined quantitative effects of climate and phenology on yield and its components. Based on data from a sowing date study in Australia (Latitude 34°04' S), the authors have developed a simple predictive model for these effects in cv. Lee. The model predicted large non-linear effects of the duration of the preflowering period on yield. These were due to direct effects of the time available for vegetative growth and to the subsequent timing of the later growth periods. To verify the predicted direct effects under 'controlled' conditions, a further experiment was conducted using photoperiodic manipulation of the vegetative period.

'Harosoy,' an early cultivar at this latitude was sown in the field on three dates in a randomized block trial with four replications. The plants were prevented from flowering by the imposition of a 24-hr photoperiod regime, using an array of mercury vapor lamps giving an illumination of 20 fc at the crop canopy. The lamps were switched off when the last sown plants had reached the first trifoliate leaf stage. Plants at all sowing dates then flowered almost simultaneously.

Yield components (total node number, pod bearing nodes, bean number per node, beans per square meter and bean growth rate) were accurately predicted by the model with the exception of the datum for bean number per node at the late sowing. The components of yield affected by the treatments were, as the model predicted, the number of nodes and hence the number of beans. There was no effect of canopy age on the gross photosynthetic capacity of the upper canopy layers, as measured by the method of Shimshi.

There was no effect of partial pod removal on bean growth during the filling phase, which suggests that the maximum bean growth rate was determined before this phase, possibly during flowering or soon after, and that the assimilate supply was sufficient to allow full expression of this maximum rate. It is concluded that it is possible to model soybean yield component responses to environmental and phenological variables with a very simple model. This type of model may have potential for assisting in soybean breeding and selection by estimating and removing effects of differing phenology.

CONTRIBUTION OF NITRATE AND DINITROGEN TO TOTAL PLANT NITROGEN AND SEED PROTEIN IN SOYBEANS

J. R. Anderson, Jr., J. E. Harper and R. H. Hageman, Department of Agronomy and USDA, SEA/FR, University of Illinois, Urbana, Illinois 61801

Nodulated soybean [*Glycine max* (L.) Merr. cv. Wells] plants were grown to maturity in outdoor hydroponic gravel culture systems. The plants were, until the R4 growth stage, subirrigated six times daily with a complete nutrient solution containing 3mM nitrate (enriched to 1.23 atom percent ^{15}N) as the sole nitrogen source. Thereafter, plants either continued to receive the same nutrient solution or were provided with a comparable solution lacking nitrate. At plant maturity, ^{15}N analyses of plant parts via emission spectrometry indicated that plants supplied with 3mM nitrate for the entire growing season obtained 69.4 percent of their nitrogen from nitrate and 30.6 percent from dinitrogen. When nitrate was removed from the nutrient solution at the R4 growth stage, nitrate and dinitrogen accounted for 36.02 and 64.0 percent of total plant nitrogen at maturity, respectively. Acetylene reduction rates and nodule fresh weights were greater in those plants that were not supplied with nitrate after the R4 growth stage. Dry matter production and seed yields from both treatments were similar. The ^{15}N analyses indicated that the contribution of nitrate to seed nitrogen was, on a percentage basis, greatest in seed from nodes at the base of the plant. It follows that the contribution of dinitrogen to seed nitrogen was, on a percentage basis, greatest in seed from upper regions of the canopy. The data emphasize the importance of nitrate to early vegetative and seed growth. Moreover, they suggest that the ^{15}N techniques utilized in this investigation may be useful in characterizing: 1) varietal differences in the contribution of the respective nitrogen assimilation systems to various nitrogen fractions in the plant, and 2) the roles of the two assimilation systems in the redistribution of nitrogen from vegetative to reproductive tissues.

DEVELOPMENTAL CHANGES IN THE REPRODUCTIVE STRUCTURES OF TWO SOYBEAN GENOTYPES

L. H. Aung, J. M. Byrne, K. E. Crosby and G. R. Buss, Virginia Polytechnic Institute and State University, Blacksburg, Virginia 24061

The developmental characteristics of a high-yielding 'Essex' and a lower-yielding 'Shore' soybean were determined. Histological examination of ovaries prior to anthesis and early pod indicated the embryological stages of mature embryo-sac, zygote, proembryo and embryo of 'Essex' were recognized and corresponded with selected advancing growth stages. In contrast, the embryological stages of 'Shore' were condensed. Both cultivars showed a high starch content in the embryo-sac prior to fertilization, but decline thereafter.

The pattern of ovary growth of 'Essex' and 'Shore' was similar, but the development of 'Essex' ovules was more advanced at intermediate growth stages than 'Shore.' Tris-soluble protein of the cultivars was similar and showed a dramatic increase in the ovules subsequent to the attainment of the pericarp size. *In vitro* culture of 'Essex' and 'Shore' showed that the cotyledonary tissues of 'Shore' responses more to 6-benzylaminopurine (BAP) than 'Essex.' Exogenous BAP treatment of developing pods of field-grown plants gave similar results.

EVIDENCE OF EXTENSIVE LIPOXYGENASE ACTIVITY DURING IMBIBITION OF SOYBEAN SEED PARTICLES

D. J. Parrish, Department of Agronomy, Virginia Polytechnic Institute and State University, Blacksburg, Virginia

Imbibition of soybeans can be very stressful. Low moisture seeds which are rapidly imbibed or subjected to chilling temperatures during imbibition show evidence of physiological injury. It has been suggested that the injury may be due to an uncontrolled mixing and/or loss of intracellular components. Such might occur during the rehydration of membranes and their reorientation into the typical bilayer configuration. Lipoxygenase (LOX) which catalyzes the oxygenation of fatty acids such as linoleate (LA) is abundant in soybean seeds, as are its substrates. The possibility exists, therefore, that LOX might, during imbibition, catalyze in an uncontrolled, non-homeostatic fashion the destruction of vital membrane components and/or produce undesirable peroxides.

Seeds of "Wayne" soybean were ground and particles 0.5 to 1 mm were separated by sieves. One hundred mg samples were placed in a 1.5-ml water-jacketed reaction vessel and their respiration at 25 C measured with an O_2 electrode. The imbibing medium was either distilled water, 10 mM phosphate buffer adjusted to various pHs, buffered solutions of various inhibitors, or buffered LA (dissolved 1:1 in Tween 20). The initial respiration of soybean particles (that occurring in the first two to three minutes after wetting) is thought to be primarily due to LOX activity because it is: 1) insensitive to 5 mM KCN; 2) 50 percent inhibited by 100 μM propyl gallate, a hydroquinone; 3) 40 percent inhibited by 1 mM salicylhydroxamate which similarly inhibits purified LOX; 4) pH dependent with an optimum about 8.0; and 5) promoted 9 to 10-fold by the addition of 50 μM LA.

These data suggest LOX activity is substantial in this imbibing system and that substrate becomes limiting within a few minutes. Obtaining data from whole, uninjured seeds is complicated by technical problems which will be discussed, but there appear to be similar rates and amounts of respiration in whole cotyledons when differences in imbibition rates are taken into account.

SOME PROBLEMS ON DETERMINATION OF PROTEIN CONTENT IN RIPENING SOYBEANS BY THE BIURET METHOD

S. Konno, National Institute of Agricultural Sciences, Kitamoto, Saitama, Japan

The changing pattern of protein content in soybean seeds with the ripening obtained by the biuret method was not always in agreement with the results taken by other methods, though very similar results were repeatedly obtained by the biuret method. Since it was considered that some problems might be involved in the application of the biuret method to protein determination of young soybean seeds, some application conditions of this method were investigated and compared with other methods.

In an experiment in which a known quantity of purified protein was added to young seed samples with biuret reagent, an inhibiting factor on biuret color development in the young seed samples could not be found. The influence of sample drying temperature on biuret values was investigated using seeds at fairly early maturing, at the height of seed filling, and at maturity of three varieties, i.e. Saikai 20, Wasemidori, and Nordin 2 were dried at 90, 70, and 50 C before being ground to powder.

In all the varieties, the samples dried at 90 C showed a lower optical density except seeds at maturity which were nearly the same as that of other temperatures in Saikai 20, while in the other varieties the density was still lower than that dried at lower temperatures. Most of the samples dried at 70 C showed slightly lower density than that of 50 C, while some samples were the same as 50 C. When the samples which were ground after being dried at 50 C were heated again at 90 and 70 C their biuret values were not so different except that of 90 C, which showed fluctuating values. The biuret values taken from 50 C dried samples correlated closely with Kjeldahl values of the samples precipitated with 0.8 N trichloroacetic acid and with the Barnstein method.

From these results, one may conclude that the samples dried at higher temperatues, especially young seeds at high moisture content are unsuitable for application of the biuret method for portein determination. Samples should be dried at 50 C and below for application of the biuret method. The biuret method is difficult to apply to soybeans having black seed coats, because the pigment (anthocyanin) is extracted into the biuret reagent and increases in brown color and optical density.

DETERMINATION OF MOISTURE IN GRAIN BY AUTOMATIC KARL FISCHER TITRATION

F. E. Jones, National Engineering Laboratory, National Bureau of Standards, Washington, D.C.

The Karl Fischer titration method has been the most widely applicable procedure for the determination of moisture in organic and inorganic materials. The method, involving the quantitative reaction of water (extracted from the material) with iodine in the Karl Fischer reagent, was described by Fischer in 1935, and was applied to the determination of moisture in grain as early as 1945. However, although the method is specific for water and was used to test the accuracy of official oven methods, it has not been adopted as the primary reference method due, at least in part, to the various disadvantages attributed to it. These included the considerable technical skill and experience required to obtain accurate, reliable results. The development of commercially-available automatic Karl Fischer titrators has overcome disadvantages attributed to the method.

In the work to be reported here, techniques have been developed to apply the automatic titrators to the determination of moisture in grain. The uncertainty in the results of measurements is typically 0.06 percent moisture content, (1 standard deviation) on a wet basis, at the 15 percent level. The particular applicability to soybeans, for which a satisfactory air-oven method has not been developed, will be shown. The Karl Fischer method is capable of determining the true moisture content of grain and, in so doing, could be used to define moisture content.

RESPONSE OF SOYBEANS TO SOIL MOISTURE STRESS

B. P. Singh, Fort Valley State College, Fort Valley, Georgia 31030

Periods of drought are quite frequent during the soybean [*Glycine max* (L.) Merr.] growing season in the southeastern United States. Shortage of water may affect plant physiological processes as well as yield. This study was designed to determine the effect of moisture stress on the nitrogenase activity, diffusive resistance and yield of soybeans. The field experiments were conducted during 1976 and 1977 on a Norfolk sandy loam soil. Five soybean genotypes, namely, Bragg, Bossier, Ransom, Ga 70-192, and N66-1136 were subjected during the reproductive period either to an optimal or a stress moisture condition. The optimal moisture condition was obtained by keeping the soil at the field capacity while the stress moisture condition was created by withholding any addition of water to the soil for a 30-day period. The entry of rain water to the stress plots was prevents by spreading plastic between the rows and stitching it together along the row of plants. The nitrogenase activity was measured using the acetylene reduction assay and the diffusive resistance of adaxial and abaxial leaf surfaces was determined with a diffusive resistance meter.

The two year average of nitrogenase activity in the optimal and stress conditions were 10.0 and 2.05 μ moles ethylene per hr per plant respectively. This amounted to approximately a 400 percent difference in the nitrogenase activity between plants at optimal and stress moisture conditions. This means that moisture stress may severely reduce symbiotic nitrogen fixation and cause nitrogen shortage in the plant. No consistent variation among the genotypes for the nitrogenase activity was observed. Moisture stress caused a significant increase in the diffusive resistance of both adaxial and abaxial leaf surfaces. The adaxial and abaxial diffusive resistance under optimal moisture conditions averaged 5.34 and 3.53 sec cm^{-1} respectively as compared to 22.14 and 5.88 sec cm^{-1} under moisture stress conditions. The increase in the diffusive resistance due to moisture stress was more pronounced in the adaxial than the abaxial leaf surface. Genotypes did not differ significantly in their resistance to water vapor diffusion. Since carbon dioxide assimilated in photosynthesis must also diffuse through stomata, increase in resistance to diffusion due to moisture stress may adversely affect photosynthesis.

Soybean yield was significantly reduced by moisture stress. The yield averages for the optimum and stress moisture conditions were 3123 and 2518 kg/ha respectively. Genotypes with higher yields at the optimum moisture level showed greater reductions in yield under moisture stress. The results of this study demonstrated the importance of optimum soil moisture conditions in soybean production.

PROTEIN SYNTHESIS DURING SEED FORMATION

J. H. Cherry, Horticulture Department, Purdue University, West Lafayette, Indiana 47907

Research from this laboratory has shown that more than 70 percent of the total protein of soybean seed during development is produced during a three-week period (30-50 days after flowering). During this very active period of synthesis a number of environmental factors may affect the quantity and quality of protein. Obviously, limitations of water, nutrients and photosynthetic capacity will impair total protein synthesis and reduce seed size. While the amount of energy (ATP) and carbon skeletons may not be manipulated, that of water and nutrients may be controlled. A major concern relating to soybean yield and protein content is to know when the plant requires the greatest amount of resources and when to provide these to prevent the occurrence of stress. Since the soybean plant is producing 3 to 4 percent of its total final protein per day during the active synthetic period of synthesis, it is reasonable to believe that a few days of water or nutrient stress could greatly reduce yield.

SOYBEAN SEED VIGOR

R. W. Yaklich and M. M. Kulik, USDA, SEA/AR/NER/AMRI, Seed Research Lab, Beltsville, Maryland 20705

The relationship between field emergence and vigor tests of soybean seeds were evaluated with seven cultivars in 1975 and eight cultivars in 1976. Field plantings were made at two sites (sandy and heavy soil types) and three dates of planting (early, pre-optimum, and optimum). In the laboratory, the seed lots were evaluated using 12 vigor tests that have been shown to be applicable to vigor testing in soybean seeds.

Field emergence was significantly better in the sandy than in the heavy soils by 5.6 and 6.3 percentage points in 1975 and 1976, respectively. Correlations between emergence counts for sites and planting dates were high ($\geq .775$) and significant ($P < 0.01$) each year. However, the analysis of variance showed that in 1975, the planting date x seed lot and site x seed lot interactions by cultivar were significant, indicating that some seed lots performed differently, depending on planting date on site. Six vigor tests were chosen because of their consistent high correlations with field emergence and their suitability for use by seed testing laboratories. These vigor tests were the speed of germination, conductivity test, accelerated aging, sand bench emergence, cold test, and tetrazolium staining. An R^2 procedure, using the vigor tests, demonstrated that using three of the vigor tests for predicting field emergence was just about as good as using all six.

EFFECTS OF PLANT POPULATION ON GRAIN YIELD AND GROWTH PARAMETERS OF SOYBEANS

P. Yingchol, Department of Agronomy, Kasetsart University, Bangkok-9, Thailand

Five varieties of soybean, Bossier, Columbus, Villiams, S. J. 2 (local variety) and S. J. 4 (local variety) were planted at four different densities, i.e. 212,440 plants/ha, 279,040 plants/ha, 362,320 plants/ha and 424,160 plants/ha. In view of this experiment, it can be said conclusively: 1) Transmission ratio values of S. J. 2 and S. J. 4 are higher than the other three varieties at the period of podding; 2) Dry matter accumulation of early stages of growth did not differ among varieties, whereas differences were detected at later stages; and 3) High density seeding resulted consistently in higher LAI than low density seeding at all sampling dates.

LOW LIGHT INTENSITY (SHADING) EFFECTS ON THE YIELD COMPONENTS OF FIELD-GROWN SOYBEANS

G. M. Prine, University of Florida, Gainesville, Florida 32611

More information is needed on the effects of short periods of low light intensity on soybeans [*Glycine max* (L.) Merr.]. 'Bragg', 'Cobb' and 'Hampton 226A' soybeans were shaded in the field for 5 or 7-day periods over most of their life cycle with black plastic fabric which allowed about 25 percent light transmission. A 5 or 7-day shade period sometimes reduced seed yield over 15 percent compared to the unshaded check. The most yield reduction occurred in shade periods just prior to flowering, after flowering in pod enlargement stage, and during the seed-filling stage. Shading in the pod enlargement stage usually resulted in reduced numbers of seed per dm of row and increased seed weights. Later shading during the seed-filling stage resulted mainly in reduced seed weights and yields and little change in seed numbers. Shading just before flowering caused severe elongation of some stem internodes and resulted in greater lodging than other shade periods. This plant lodging, and not the direct effect of the shading, may have caused the reduced seed yield found for the shade period just prior to flowering.

CULTIVAR VARIATION IN THE LENGTH OF THE GRAIN FILLING PERIOD

D. A. Reicosky, Agronomy Department, University of Kentucky, Lexington, Kentucky 40506

Nine cultivars which varied in seed size, growth habit and maturity were evaluated in a field environment for variation in the length of the grain filling period (GFP). Methods of measuring the GFP which can be used efficiently in a plant breeding program were also evaluated. The range in the GFP estimated by actual measurements of the beginning and end of seed growth on several pods of 1 plant/ rep was 29 days for Fiskeby V and 47 days for Emerald. The large seeded strains had GFP's of 40 (Magna), 42 (Verde), and 47 (Emerald) days while the small seeded strains had GFP's of 33.9 (OX-303) and 41.1 (Essex) days. Essex and OX-303, both determinant varieties, had 33.9 and 41.1-day GFP's while M65-217 and Williams indeterminant varieties had 34.4 and 40.8-day GFP's. The three early varieties (Fiskeby V, M65-217 and OX-303) had shorter GFP than the later varieties (Essex and Emerald), although there was some variation in the early varieties. A larger sample of the germplasm is necessary before definite conclusions can be made about the relationship of GFP and seed size, maturity and growth habit. The range of the GFP for single pod measurements per plant at the beginning and at the end of seed growth was 29 days for Fiskeby V and 51 days for Emerald. These two methods were in agreement with a correlation coefficient of .945. The effective filling period (EFP) estimate was smaller than the GFP estimate, however there was general agreement of the techniques with a correlation coefficient of .694.

All estimates of the filling period gave comparable results and can be used to measure the filling period in soybeans. The choice of technique will depend on the objective, volume and precision required in the experimentor's design. Techniques are now available for evaluating the length of the grain filling period on an individual pod or plant basis or for estimating the GFP of larger plots. Thus this complex physiological character is now amenable to classical genetic studies and could be manipulated in a plant breeding program. The strain variation in the GFP in the limited germplasm evaluated here indicate that genetic variation does exist for the GFP. Basic information both genetic and physiological in nature is necessary to determine fully the potential of utilizing this character in cultivar improvement programs.

PHYSIOLOGICAL BASIS FOR YIELD DIFFERENCES IN SELECTED SOYBEAN CULTIVARS

S. Gay, D. B. Egli, and D. A. Reicosky, Department of Agronomy, University of Kentucky, Lexington, Kentucky 40506

Field experiments were conducted in 1976 to investigate the physiological basis for yield differences between old, low yielding and new, high yielding soybean cultivars. The low yielding cultivars used were Lincoln (maturity group III) and Dorman (maturity group V). The high yielding cultivars were Williams (maturity group III) and Essex (maturity group V). These four cultivars were grown in the field using conventional practices with supplemental irrigation. Mechanical support was used to prevent lodging for comparison with control plots allowed to lodge normally. There was no lodging in the maturity group III cultivars; however, Dorman lodged significantly more than Essex in the control plots, but there was no relationship between lodging and yield. Williams produced 34 percent more yield than Lincoln, primarily because of larger seed. Essex produced 88 percent more yield than Dorman, primarily because of more seed per unit area.

Yield differences were not correlated with total shoot weights or the CO_2-exchange rate of a single leaf during vegetative or early reproductive growth. Rates of acetylene reduction were similar for Williams and Lincoln, but Essex had a higher rate than Dorman. The yield advantage of Williams resulted mainly from a longer filling period which was associated with a higher CO_2-exchange rate and rate of acetylene reduction late in the grain filling period. Essex and Dorman produced similar total shoot weights so the yield difference apparently resulted from a more efficient partioning of photosynthate to the seed in Essex. The data suggest that yield improvement in the future may be possible by lengthening the grain filling period. However, it may not be possible to further increase the partioning of photosynthate to the seed, if the partioning in current cultivars is approaching 100 percent.

EFFECT OF TEMPERATURE ON EMERGENCE, HYPOCOTYL GROWTH AND DRY MATTER PRODUCTION IN SOYBEANS

G. F. Nsowah, University of Science and Technology, Kumasi, Ghana

Seedling emergence has been a major problem to soybean production in the tropical and subtropical regions of the world. Many factors including harvesting date, threshing methods, drying temperature, length and duration of storage, soil temperature and soil moisture at planting have been known to affect seedling emergence and hypocotyl development. Seeds of four contrasting varieties of soybean, namely Hardee, Improved Pellican, Bossier and TG 627 were sown in seed boxes filled with sterilized sand and kept under darkness in growth chambers with temperatures maintained at 25, 30, 35, 40, 45, and 49 C respectively.

Results confirm the existence of significant differences between the varieties and the different temperatures on the percentage seedling emergence and the rates of hypocotyl development. The rates of seedling emergence were faster at temperatures between 30 and 40 C than at temperatures below 30 C and above 40 C. The rates of hypocotyl development were faster at temperatures between 25 C and 35 C than those at higher temperatures up to 40 C. At 49 C there was hardly any hypocotyl development after seedling emergence.

SEED TREATMENT WITH GROWTH REGULATORS TO CONTROL HEIGHT OF GREENHOUSE GROWN SOYBEANS

D. R. Smith, D. W. Unander, O. Myers, Jr., and J. H. Yopp, Southern Illinois University, Carbondale, Illinois 62901

Soaking soybean seed in growth regulators prior to planting was effective in reducing internode length and, hence, height of greenhouse grown plants. Of the three growth regulators tested—Amo 1618, ancymidol, and ethephon—Amo 1618 gave the most satisfactory control of plant height. Height of plants from seed soaked in 250 ppm Amo 1618 averaged less than one-third that of controls at flowering. At this concentration, there was no significant reduction in percent germination. Higher concentrations of Amo 1618 or combinations of Amo 1618 and ancymidol were only slightly more effective in reducing plant height. Ancymidol alone or ethephon alone were effective only at concentrations that reduced germination. Flowering and pod set were not affected significantly by the Amo 1618 seed treatments. The seed soak method had two major advantages over foliar application of the growth regulators, the first was better control of early seedling growth and the second was that only one application rather than several was required. The Amo 1618 seed soak treatment is being used currently in the generation advancement phase of a single-seed-descent breeding program to reduce the labor requirements of maintaining greenhouse grown plants.

FLORAL INITIATION OF SOYBEANS AS GOVERNED BY PHOTOPERIOD, TEMPERATURE, AND STAGE OF DEVELOPMENT

J. F. Thomas and C. D. Raper, Jr., North Carolina State University, Raleigh, North Carolina 27650

Soybean plants of the determinate cultivar 'Ransom,' grown under 16-hr photoperiods to the second and fourth node stages, were exposed to photoperiods of 10, 12, 14, 15, and 16 hr at day/night temperatures of 18/14, 22/18, 26/22, 30/26, and 34/30 C. Photoperiod treatments included a 9-hr duration of high intensity light from a combination of fluorescent and incandescent lamps (PPFD of 67.0 nE cm^{-2} sec^{-1}). The photoperiods were extended with light from incandescent lamps only (PPFD of 8.4 nE cm^{-2} sec^{-1} and PR of 11.7 W m^{-2}). Day temperatures were coincident with the 9-hr, high intensity light period. Three plants from each treatment were sampled at 2 to 3-day intervals over a 21-day treatment period. Each axil was examined for occurrence and number of microscopically discernible floral primorida. Floral primordia were found at all photoperiods and temperatures during the 21-day sampling period; however, floral initiation was delayed as photoperiod increased and temperature decreased. The number of sites (axils) for primordia increased with both photoperiod and temperature. Anthesis had occurred by the twenty-first day only at the 10 and 12-hr photoperiods and at temperatures of 26/22 C and above. Differentiation of floral primordia continued at different nodes over several weeks, but anthesis, when it occurred, was within a few days at all nodes. Other data collected included dry matter, soluble carbohydrate, and nitrogen contents of leaves, stems, and roots; number and area of leaves; and number and lengths of branches and length of main stem. These results are being used to extend a dynamic model of soybean growth to describe reproductive development.

EFFECTS OF SHORT-TERM CHILLING ON THE GROWTH OF THREE VARIETIES OF SOYBEAN

D. T. Patterson, Southern Weed Science Laboratory, USDA, SEA/AR and Mississippi Agricultural and Forestry Experiment Station, Stoneville, Mississippi 38776

The effects of simulated naturally occurring chilling events on the growth of three varieties of soybean [*Glycine max* (L.) Merr.] were determined in controlled-environment regimes, chosen to represent early season temperature conditions typical of those occurring where the varieties are grown normally in the United States. 'Tracy' soybeans were grown at a 27/23 C day/night for 24 days, exposed to a 24/18 C day/night for 3 days, and returned to 27/23 C for an 11-day recovery period. 'Williams' soybeans were grown at 24/20 C, exposed to 17/10 C for 3 days, and returned to 24/20 C for recovery. 'Corsoy' soybeans were grown at 19/11 C, exposed to 11/4 C for 3 days, and returned to 19/11 C. Plants of the same varieties were also maintained continuously in the warmer temperature regimes to serve as controls. Harvests were made at the beginning and end of the 3-day cold period and at the end of the recovery period. Leaf areas and total dry weights were determined at each harvest. Mathematical growth analysis techniques were used to calculate net assimilation rates (NAR = g $dm^{-2}day^{-1}$), leaf area durations (LAD = dm^{-2}days), and dry matter production (ΔW = NAR x LAD) over each harvest interval.

The 3-day chilling treatment reduced significantly the leaf area production by 15 to 20 percent in all three varieties. The ΔW during the 3-day chilling period was not affected in 'Williams', but was reduced significantly in 'Corsoy' and 'Tracy'. The reductions in ΔW in 'Corsoy' and 'Tracy' were caused by reductions in both NAR and LAD during the cold treatment. At the end of the recovery period, both dry weights and leaf areas were significantly less in the cold-treated plants than in the controls. The reductions in ΔW in the cold-treated plants during the recovery period were caused primarily by their lower LADs.

In a second experiment, the effect of chilling on photosynthesis in 'Corsoy' and 'Tracy' soybeans grown at 28/21 C and exposed to 17/9 C for 3 days was determined at a measurement temperature of 28 C. The average photosynthetic rate of 'Tracy' soybeans exposed to the chilling treatment was 17.7 mg CO_2 $dm^{-2}hr^{-1}$, significantly lower than the rate of 26.7 mg CO_2 $dm^{-2}hr^{-1}$ for the unchilled controls. In 'Corsoy', chilling reduced significantly the photosynthetic rate from 21.2 to 11.4 mg CO_2 $dm^{-2}hr^{-1}$. Therefore, the percentage reduction in photosynthetic rate due to chilling was greater in 'Corsoy' (46 percent) than in 'Tracy' (34 percent).

KINETICS OF FATTY-ACID CHANGES IN SOYBEAN ROOT SYSTEMS IN RESPONSE TO A CHANGE IN GROWTH TEMPERATURE

A. H. Markhart, III, and P. J. Kramer, Department of Botany, Duke University, Durham, North Carolina 27706

Although the ability to acclimate to a change in temperature is an important adaptive characteristic to plants growing in a fluctuating environment, little is known about the speed or kinetics with which these changes occur. The experiments reported here examine the rate at which fatty acid composition of soybean (var. Ransom) changes in response to a change in temperature. Osmotic potential, water potential and stomatal resistance were also measured. Plants were grown in the Duke University Phytotron in a thermoperiod of 29/23 C and a 16-hr photoperiod. After 3 weeks of growth, one-half of the plants were transferred to a 17/11 C thermoperiod. On 1, 2, 4, and 8 days after transfer, 2 plants were harvested for GLC fatty acid analysis. Water and osmotic potential measurements were made in triplicate on leaf discs and stomatal resistances were measured between 2 and 3 p.m. each day. Control plants were also grown at 17/11 C from seed. Results presented are the average of 2 experiments.

Preliminary experiments showed that a significant increase in the fatty acid unsaturation of plants grown at 17/11 C compared to 29/23 C. Over 95 percent of the change was due to change in the relative amounts of linolenic (18:3) and linoleic (18:2) acid. In the 29/23 C grown plants the 18:3/18:2 ratio averaged 1.02 whereas in the 17/11 C plants the ratio averaged 1.3. When plants were moved from the 29/23 C to the 17/11 C environment the change in fatty acid composition began immediately, with half the change occurring in 2 days and 95 percent after 8 days. Unfortunately, our procedure does not discriminate between alterations in mature root membranes and the changes due to new growth.

The plants transferred to the 17/11 C temperature showed no visible signs of water stress. Turgor potentials were always as high as plants in the 29/23 C treatment. There was, however, a significant decrease in the water and osmotic potential that was apparent 2 days after transfer, and continued through the 8 days of the experiment. In the warm treatment, water potentials and osmotic potentials averaged around -9 bars and -11.5 bars respectively throughout the experiment, whereas the cold plants decreased to -12.4 and -14.1 bars after 2 days and -15.5 and -18.5 after 8 days. There were no significant differences in stomatal resistances between the 29/23 C grown plants, the 17/11 C grown plants, or the ones that were transferred from the warm conditions to the cold.

In 'Ransom' soybeans, an increase in fatty acid unsaturation is part of the plants acclimation response to lower temperatures. The change begins immediately, is essentially complete in 8 days, and involves primarily a shift in the C18:3 and C18:2 fatty acids ratio. The initial shock of the temperature change had no dramatic effects on stomatal resistance. By the second day, however, both the osmotic and water potentials of the cold exposed plants had decreased significantly compared to the controls. This trend continued for the eight days of the experiment. The rate at which a plant can acclimate to a change in environmental condition such as an abrupt lowering of temperature may be an important characteristic in terms of survival and productivity. Further efforts along those lines of investigation are warranted.

RELATIONSHIP OF SOIL MOISTURE TO GROWTH AND NITROGEN ASSIMILATION BY FIELD-GROWN SOYBEANS

A. B. Tambi, R. P. Patterson, H. D. Gross, and C. D. Raper, Jr., North Carolina State University, Raleigh, North Carolina 27650

Three soybean [*Glycine max* (L.) Merr.] cultivars, representing maturity groups V (Forrest), VI (Lee 74), and VII (Ransom), were grown at three levels of soil water adjustment in the field in 1977 and 1978. Moisture treatments were 1) sprinkler irrigation supplementing rainfall, 2) rainfall as the only source of water, and 3) plant canopy covered with 4 ml clear plastic sheets (reduced light by about 17 percent) as shelters to induce soil and plant water stress from first flowering through about the mid-pod-fill stage. Soil and plant water stress determinations (using gravimetric and leaf water potential thermocouple psychrometric methods, respectively), acetylene reduction of excised nodulated root systems (to estimate nitrogenase activity), nitrate reductase (NR) activity of the middle leaflet of the third trifoliolate leaf, and dry weights of all plant parts were measured at weekly intervals from about three weeks after planting through physiological maturity. For both 1977 and 1978, seed and pod yield response was greatest for Forrest, and lower for Lee 74. Ransom did not respond to supplemental irrigation. Total vegetative dry weight of Ransom exceeded consistently that of Forrest and Lee 74 under all treatment conditions, but there · as little difference within Ransom under these treatments. Both total and specific nodule activity and NR activity were reduced under water stress. Under water stress, specific nodule activity for Ransom was the same as that for Forrest. Ransom NR activity, however, was higher than that for Forrest. An important implication of this work is that supplying adequate moisture during early pod fill extends the capacity for nitrogen assimilation, via nodule function and nitrate uptake and reduction, to later stages of embryonic development when the plants otherwise would be water-stressed.

PRODUCTION

Invited Papers

FOLIAR FERTILIZATION OF SOYBEANS

J. J. Hanway, Iowa State University, Ames, Iowa 50011

Foliar application of fertilizer nutrients during seed-filling appears to be a promising method for increasing soybean yields. During seed-filling the supply of soluble carbohydrates going to the plant roots is limited resulting in reduced root growth and activity and reduced N-fixation in the nodules. Because of the high nutrient requirements of the developing seeds nutrients are translocated from the leaves and other vegetative plant parts to the developing seeds resulting in nutrient depletion in the vegetative plant parts. Even so, the nutrient supply to the developing seeds usually is inadequate and limits seed yields. Therefore, foliar application of needed nutrients can result in very significant yield increases.

Field experimental results to date show that yields have been increased by foliar application of fertilizer nutrients during seed-filling in several field experiments. But, in many other experiments there was no yield increase and, in some cases, yields were decreased. Obviously, we still have things to learn.

The experiments where foliar fertilization during seed-filling did increase yields show: 1) The foliar application must include N, P, K and S. These nutrients must be present in a ratio of approximately 10:1:3:0.5 of N:P:K:S similar to the ratio of these elements in the soybean seed. 2) The yield increases generally have resulted from an increase in number of seeds harvested, not an increase in seed-size. 3) Foliar fertilizer applications should be made in the evening or early morning. Midday applications when the sun is shining often result in severe leaf "burn." 4) Urea is more effective and less toxic than other common forms of N. But, the rate of urea-N per spraying generally should not exceed 20 kg N/ha to avoid leaf "burn." 5) Best yield increases have resulted from 3 or 4 sprayings during seed-filling.

Studies are being conducted to try to solve the problems associated with foliar fertilization. Since intermediate break-down products of urea such as biuret have been shown to be toxic to plant roots, the toxicity of different break-down products in foliar applications is being studied. Preliminary results indicate that cyanate and cyanamid are more toxic than biuret. Carbonate at similar rates of application were not toxic. The possible general need for other nutrient elements (such as Fe, B, Ca, Mg, Mn, Cu and Zn) in addition to NPKS is being studied. Experiments are being conducted to study foliar fertilization as influenced by different varieties, row spacings, soil moisture levels, solution pH, adjuvants and other variables. There is much to be learned. But, the potential for increasing crop yields and increasing fertilizer use efficiency is so great that we must continue these studies and try to solve the problems.

FACTORS AFFECTING THE RESPONSE OF SOYBEANS TO MOLYBDENUM APPLICATION

F. C. Boswell, University of Georgia, Georgia Station, Experiment, Georgia 30212

Interest in the use of the micronutrient molybdenum (Mo) for soybeans, *Glycine max* (L.) Merr. has been stimulated in recent years because of significant yield responses. These responses occur where Mo is deficient in the soil or availability is reduced because of various environmental factors. Molybdenum is a particularly important micronutrient for soybeans because of the essential functions it has in symbiotic nitrogen fixation and nitrate reduction processes in the plant.

Soybean yield response to supplemental Mo application has been reported in the Far East (Japan, China, Taiwan), Europe, and at least 10 states of the United States. Generally, the most frequently observed responses are in the Eastern U.S. where rainfall is moderate to heavy and soils tend to be acid. Yield increases of more than 65 percent from the application of Mo to soybean seed, as a foliar spray or as soil applications, have been reported. Responses have been obtained with seed applied Mo rates as low as 35 g/ha. Also, molybdenum applications often increase seed protein and decrease percent oil.

Major factors that affect soybean response to Mo applications are low soil Mo levels, acid soil conditions and low plant uptake of Mo. Total soil Mo usually ranges from less than 0.5 to 4.0 ppm with extremely high Mo soils having 5 to 10 times these levels. Only a small portion of this total Mo is available for plant uptake. In general, acid ammonium oxalate extractable Mo is below 0.2 ppm for soils that respond to supplemental Mo at low soil pH values. Soybean leaf tissue often has Mo concentrations of less than 0.1 ppm when grown in deficient soils while typical plant concentrations are 0.5 to 2.0 ppm. Fruiting plant parts (soybean seed) often contain 2 to 25 times as much Mo as tissue. Various factors that may influence the response of soybeans to molybdenum application will be discussed.

TILLAGE SYSTEMS AND EQUIPMENT FOR SOYBEAN PRODUCTION

W. G. Lovely, Iowa State University, Ames, Iowa 50011

Abstract not available at press time.

SOYBEAN HARVESTING EQUIPMENT: RECENT INNOVATIONS AND CURRENT STATUS

W. R. Nave, SEA/AR, USDA, Urbana, Illinois 61801

Harvesting is one of the most critical steps in profitable soybean production. Surveys as late as 1971 indicated that the average soybean producer was losing 8 to 10 percent of his soybean crop in the harvesting operation. A concentrated research and development program by both public researchers and engineers from industry has resulted in the development of equipment which reduces harvesting losses to less than 3 percent of the crop yield.

The value of using floating cutterbar attachments has been known for several years. More recent innovations include the use of air-jet guards, row units with a pulling device and a row crop header with rotary cutters. Results from tests of these improved harvesting devices will be discussed in some detail.

With the introduction of combines equipped with rotary threshing devices, studies have been initiated at several universities to determine the value of rotary threshing in reducing soybean damage. Tests at Ohio State University in 1976 and at the University of Illinois in 1977 indicate that soybean damage as measured by percent splits was significantly higher for a conventional cylinder than for rotary threshing mechanisms. However, when the conventional cylinder was operated within the manufacturer's recommended cylinder speed, the splits from threshing did not exceed the allowable limit for USDA No. 1 grade soybeans.

Automatic controls to monitor the machine and crop status and perform the necessary adjustments previously required of the operator may become an integral part of future harvesting equipment. For several years grain loss and shaft speed monitors have been used to alert the combine operator to excessive threshing and separating losses. Research at the University of Illinois in 1978 indicated that an automatic control system can be used to optimize combine cylinder speed as a function of grain moisture. Research has been initiated at Clemson University to develop automatic control systems to monitor several combine functions and optimize the system based on pre-determined harvesting parameters.

CRITICAL FACTORS IN SOYBEAN EMERGENCE

H. D. Bowen, North Carolina State University, Raleigh, North Carolina 27650 and J. W. Hummell, SEA/FR, University of Illinois, Urbana, Illinois 61801

Emergence of viable soybean seedlings is very rapid in the absence of limiting environmental stresses. However in the event one or more of the edaphic factors—temperature, moisture, aeration or mechanical impedence—becomes limiting, or seed vigor is below par, all aspects of the planting operation and some aspects of the disease environment become critical. The objectives of this paper are to put numbers on the stress tolerance of germinating soybean seedlings to limiting aspects of the physical environment and relate these to the stress factors caused by weather-soil interactions, the presence or absence of pathogens, the soil physical condition, the time of planting, and planter adjustments. Field results for three growing seasons showing emergence and yield response to planting depth, seeding rate and fungicide treatment are discussed.

Contributed Papers

LIMITED SEEDBED PREPARATION SCHEMES FOR PLANTING SOYBEANS

J. H. Palmer and T. H. Garner, Clemson University, Clemson, South Carolina

The intensive cropping of soybeans and corn on many Coastal Plains fields in the Southeast has increased the incidence of soil erosion. Coupled with this have been increased costs to the farmer for labor, equipment, and energy. One of the major culprits with both increased erosion and costs is the tillage for seedbed preparation.

In 1976-78, experiments were conducted evaluating various systems of reduced tillage in preparing a seedbed for soybeans. These systems referred to as limited-seedbed systems involved tillage and planting equipment such as: the "Super-Seeder"—no-til planter; the Lely Roterra; the power-driven rotary hoe; and an experimental once-over planter designed by the Clemson University Agricultural Engineering Department. The objective was to study stand, weed control, yields, and costs for several limited-seedbed and conventional seedbed preparation methods for the Coastal Plains.

The results from the three-year study indicate: 1) deep tillage, especially row subsoiling, was a necessary component of any limited seedbed system for a typical light-textured Coastal Plains' soil; 2) several systems studied gave yields, stand, and weed control comparable to conventional methods; 3) reduced costs, erosion, and time lost were possible for several systems; 4) with some systems, increased trash (crop stubble and weeds) made weed control by the preplant herbicides (e.g. Treflan) more difficult; and 5) more intensive management, i.e., increased seeding rates, more dependence on herbicides, closer scrutiny of equipment, etc., was needed for a successful limited seedbed system for soybeans. To summarize, many Coastal Plains' farmers may be able to conserve soil and water, plus reduce costs by reducing tillage and the number of trips across the field in planting soybeans.

BROADCAST SEEDED, MONOCULTURED AND DOUBLE CROPPED SOYBEANS ON HEAVY SOILS IN MISSISSIPPI

F. D. Whisler, N. W. Buehring, and J. K. Young, Mississippi Agricultural and Forestry Experiment Station, Mississippi State, Mississippi

In Mississippi there is a high probability of extended rainfall during the recommended planting dates of May 15-30. During late March and early April, there is a lower probability of rainfall; thus in monocultured soybeans, a seedbed could be prepared in early spring. Late planted soybeans, after June 15, are often subjected to yield reductions due to August and September water stresses. This research was designed to test the feasibility of broadcast seeding of soybeans (simulated aerial seeding) in either a prepared seedbed or standing wheat during the recommended planting date and to compare the results to conventionally planted soybeans in either monocultured or double cropped conditions.

Starting in 1975 two soybean cultivars, Tracy and Forrest, were planted on a Tuscumbia clay loam soil at the Northeast Mississippi Branch Experiment Station. Eight planting methods, including six double cropping and two monocultures, were tested. The conventional system was to disk, chisel incorporate herbicides with a spring-toothed harrow and plant in 120-cm rows. During the course of this study, percent emergence, stand, yield, visual weed ratings, soil moisture, soil bulk density and soil penetration resistance have all been monitored.

The results indicate that in those seasons with high rainfall during the recommended planting time that broadcast seeded soybeans yielded as well as conventionally planted soybeans in both monocultured and double cropped systems. If surface soil moisture was low and the soil hard, as in 1977, the broadcast soybeans did not emerge. They had to be replanted with a drill. This led to a delay in soybean development and yield reductions. Broadcast seeding of wheat into soybeans and soybeans into wheat presents a new set of problems in weed control and would not be recommended on a continuing basis. For those seasons however when the soil is wet during the recommended planting date, aerial seeding of soybeans gives as good a yield as conventionally planted beans and better than late, conventional planted soybeans following wheat.

EFFECT OF VARIOUS PLANTING PRACTICES ON SOYBEAN PRODUCTION IN MICHIGAN

Z. R. Helsel, T. J. Johnston, and D. Merck, Michigan State University, East Lansing, Michigan

Soybeans are a relatively new crop to Michigan. Being virtually surrounded by lakes, Michigan experiences different environmental conditions than other northern states. A study was initiated in 1978 to evaluate the effect of various agronomic planting practices on soybean production in Michigan. The experimental sites were located in central and southern Michigan. Four cultivars, 'Evans' (early maturity), 'Hodgson' (medium maturity), 'Corsoy' (late maturity), and 'SRF 200' (narrow leaf type) were grown. These varieties were planted at two populations (344,177 and 473,244 plants/ha) in three row spacings (25, 50, and 75 cm), on three planting dates (May 8 and 23 and June 7 for central Michigan; May 25, June 15, July 7 for southern Michigan). Measurements included yield, maturity, lodging, and several components of yield. A split, split plot with three replications was used for the experimental design.

Yields in central Michigan (25.5 q/ha) were greater than in southern Michigan (20.4 q/ha). Yield reductions in southern Michigan resulted from delayed planting due to wet soils and a mid-summer drought. The drought also led to a hastened maturity and reduced plant height in southern Michigan. Later dates of planting resulted in significant ($P<.05$) yield decreases except in central Michigan where no difference was found between the first and second planting date. Cool soil temperatures slowed emergence of the first planting narrowing the difference in emergence dates of the two plantings. The 25 and 50-cm row spacings resulted in significantly ($P<.05$) greater yields than 75-cm rows in both locations. Population did not affect yield in either location. 'Evans' the early maturing cultivar in these studies, yielded significantly ($P<.05$) less than other cultivars. Dates of maturity stage R5 occurred in central Michigan on August 2, 6, and 13 for early, normal, and late plantings respectively (averaged over all other variables). In southern Michigan R5 occurred on August 5 and 15, and September 2 respectively for the three planting dates. Dates of maturity also showed that late planted soybeans progressed through stages of development more rapidly than earlier planted soybeans.

Lodging did not differ greatly among the various planting practices in southern Michigan because the late planting dates and dry weather resulted in shorter than normal plants. In central Michigan the earlier maturity cultivars showed less lodging. The 75-cm rows and high population resulted in more lodging than the narrower rows or low population. Continued research will be needed to establish characteristic trends of soybean growth and production in response to planting alternatives.

EFFECT OF PLANTING DATE AND ROW SPACING ON SOYBEAN YIELD IN THE RAINFOREST ZONE OF GHANA

H. Mercer-Quarshie, Crops Research Institute, Kumasi, Ghana

There is a resurgence of interest in the cultivation of soybean after a break of many years in the promotion of the crop. Proper management practices need to be sought and adopted if cultivation of soybeans is to be profitable. Therefore in 1974, 1975 and 1976 the effect of row width and planting date on yield was investigated. A double split plot design, with varieties Davis and Jupiter as main plots, row widths (20, 35, 50 and 65 cm) as subplots, and 5 two-weekly planting dates (beginning in April) as sub-subplots, was employed. There was a reduction in yield with increase in row width but significant differences in yield due to row width were obtained only between 20 and 65 cm rows and this only in 1974 and 1976. In 1974 when there was no break in the rains in August (usually a dry month), yield of Jupiter was reduced with later than mid-April plantings while that of Davis was increased with delayed plantings. In 1975 and 1976 when the usual dry August was experienced, plantings of Jupiter after April resulted in reduced yields. Davis behaved in a similar manner except that its optimum planting date occurred two weeks after that of Jupiter. The trend of decreasing yield with late planting was associated closely with reduced rainfall during the pod filling period of the crop.

INFLUENCE OF SOIL TEMPERATURE AND DEPTH OF PLANTING ON GERMINATION AND EMERGENCE OF TROPICAL AND TEMPERATE SOYBEAN CULTIVARS

H. J. Hill, J. Oard and W. H. Judy, INTSOY, Department of Agriculture, University of Illinois, Urbana, Illinois 61801

Cultivars which can emerge quickly from the soil when temperatures are sub-optimal or can emerge from a deep planting have a greater competitive advantage than ones which do not. More importantly, cultivars which have this ability under these stress conditions may possess a greater seedling vigor than others independent of seed quality. The objectives of this study were to test the ability of a wide range of cultivars to germinate and emerge under low, moderate and high temperatures, and to emerge from moderate and deep plantings. The 36 cultivars chosen are currently entered in ISVEX and SPOT. The same seed lots were used for all three experiments planted. Planting depths were five and ten cm. Soil temperatures were altered by planting at different locations and different seasons. Clear polyester plastic was used on half the plots in two tests to artificially raise temperatures. Mean soil temperatures for the first week after planting were 28, 24.5 and 12 C for the three tests at the five-cm depth. Mean daily temperatures ranged from 7.2 to 30 C and hourly temperatures ranged from 6 to 35 C. Soil moisture was monitored and found not to be limiting.

In all three tests cultivars differed significantly in emergence rates. Cultivars also differed significantly in total emergence from the ten-cm depth. There was significant cultivar x depth interaction only in the low temperature test. Emergence rates were highest at the high temperature test and lowest at the low temperature test. In general, cultivars which had a high total emergence from the cold temperature test possessed high emergence rates in the moderate and high temperature test. Likewise, cultivars which had low emergence rates from the high temperature test had low total emergence under the cold temperature test. The tropical varieties used tended to have lower emergence rates than temperate varieties under the high temperature test. Under the cold temperature test the cultivars with the higher emergence rates were mostly developed in the northern United States. Whether this is related to the quality of the seed lot used or is inherent in the variety will be discussed.

INFLUENCE OF ROW-WIDTH ON SOYBEAN CULTIVARS UNDER TROPICAL AND TEMPERATE ENVIRONMENTS

J. Oard, H. J. Hill and W. H. Judy, INTSOY, Department of Agronomy, University of Illinois, Urbana, Illinois 61801

This study was designed to investigate the influences of decreasing row width on cultivar performance. Two experiments each consisting of 16 varieties were planted at Isabela, Puerto Rico, in July. The varieties represented maturity groups III - IX. A third experiment was planted at Urbana, Illinois, during May so that comparisons may be made on the influence of row width under tropical versus temperate environments. The row widths used were 20, 40 and 60 cm. These were constant in every experiment. The populations were nearly constant among experiments with approximately 400,000 plants/ha at Urbana and 440,000 plants/ha for the two experiments in Puerto Rico. Many varieties were common to two tests and one variety common to all three.

Among results common to all three tests lodging was found to increase with row spacing at these high populations. Plant height increased with decreasing row width providing that the variety had a normal vegetative growth period. Date of flowering and maturity were not affected by row width. Grain yield kg/ha) was increased with decreasing row width in Urbana for the earlier maturing varieties but not for the later maturing varieties. In Puerto Rico, an increase in grain yield occurred with decreasing row width among the earlier maturing varieties also. However, for maximum benefit the variety had to undergo a normal vegetative growth stage to reap full benefits.

MULTIPLE CROPPING OF SOYBEANS WITH SMALL GRAINS IN SOUTHERN WISCONSIN

S. L. Kaplan and M. A. Brinkman, Department of Agronomy, University of Wisconsin, Madison, Wisconsin 53706

Multiple cropping, production of more than one crop from a field in one growing season, may be accomplished in several ways. Crops may be grown in succession (e.g., douple cropping) or they may be grown such that a second crop is seeded into a previously established crop with each crop harvested separately (i.e., relay cropping). The primary advantage of such systems may include increased productivity of a field. The purpose of this research is to compare relay and double cropping systems with soybeans (*Glycine max* L. Merr., 'Hodgson') and either spring oats (*Avena sativa* L., 'Lang') or spring barley (*Hordeum vulgare* L., 'Beacon' and X1890-2). Treatments were devised so that the effects of various competition factors, such as row spacing, planting date, and small grain harvest as silage or mature grain, could be determined. In 1977, the small grains were drilled in a split plot arrangement of a randomized complete block design of four blocks and with the small grain genotypes as the whole plots. Soybean row spacings of 45, 75, and 90 cm were combined, respectively, with gaps created in the small grain rows of 30, 45, and 60 cm in which soybeans were seeded. In 1978, the experiment was repeated with the addition of such variables as two seeding rates of Lang oats and the harvest of Beacon barley as high moisture grain in addition to silage and mature grain. All variables were combined factorially except for such restrictions as double cropping of soybeans following the harvest of the small grains only for silage.

The soybean checks yielded 3470, 3280, and 2420 kg ha^{-1} in the 45, 75, and 90-cm row spacings, respectively. The closest soybean yields in any multiple cropping system were with Lang oats when harvested as silage in either a relay cropping system (2010, 1960, and 2010 kg ha^{-1} in the 30 to 45, 45 to 75, and 60 to 90 cm gap-row spacing combinations) or a double cropping system (2150, 1940, and 1990 kg ha^{-1}). The silage yields in these treatments were also very similar, with the relay cropping system producing 3120, 2770, and 2400 kg of dry matter ha^{-1} and the double cropping system producing 3160, 2780, and 2470 kg ha^{-1}. One of the most significant observations to be made in the above data is the similarity in the productivity of the two multiple cropping systems where the highest soybean yields were attained. Apparently the normal yield advantage of the earlier soybean planting data in the relay cropping system was negated by the competition of the oats during early development of the soybeans.

SOYBEAN STAND ESTABLISHMENT AND YIELD AS AFFECTED BY HERBICIDES AND PLANTING PRACTICES

R. R. Johnson, University of Illinois, and L. M. Wax, SEA/USDA, Urbana, Illinois 61801

When reduced soybean [*Glycine max* (L.) Merr.] stands occur, herbicides are often blamed for the problem. Our objective was to determine if metribuzin and/or vernolate herbicides interact with other cultural variables resulting in stand reductions. Cultural variables researched included soybean cultivar, seed quality, planting depth, planting date and surface versus incorporated herbicide application. In all experiments the herbicides were applied at slightly higher than recommended rates, and soybeans were planted in 75 cm row widths at a density of 350,000 seeds/ha.

In several cases visual symptoms of herbicide leaf damage occurred, but the herbicide had only minor effects on stand establishment and grain yield. However, several other variables significantly reduced stands and grain yield. For example, differences in seedbed condition created by different planting dates altered stand establishment. Compared to a 1.9 cm planting depth, a 3.8 cm planting depth reduced stands in cold and crusted seedbeds. In contrast, the shallow depth caused stand problems in a dry seedbed. Seed lots with acceptable warm germination scores but decreasing vigor levels (low cost test germinations) also performed poorly in a number of instances. Of the cultural variables affecting stand establishment, none exhibited a significant interaction with herbicide treatment.

Our data suggest that cultural variables such as seedbed conditions, seed quality and minor differences in planting depth are often the major variables causing stand problems. Visual symptoms of herbicide damage that do not result in death of the plant are often outgrown in a short time. Extremely high herbicide rates such as caused by spray overlaps or carryover from a previous year may result in plant death and associated yield loss.

SOYBEAN AND CORN YIELDS UNDER DIFFERENT INTERCROPPING PATTERNS

S. Galal, L. Hindi, M. M. F. Abdalla, and A. A. Metwally, Department of Agronomy, Faculty of Agriculture, Cairo University, Giza, Egypt

The importance of soybean as a pulse crop rich in protein and oil content prompted interest in its inclusion in the crop rotation in Egypt. Due to competition with principal summer crops such as cotton, rice and maize, soybean can only be produced as an intercrop with maize. Different cultivars of soybeans and location varieties and populations of corn were intercropped under different patterns. Different characters of intercropped corn were not affected by the soybean variety. The main effects were due to the patterns of intercropping corn and soybean. Intercropping on alternating ridges increased the percentage of harvested corn plants, number of ears per plant and grain weight per ear and sometimes per plant. Grain yields of intercropped corn increased by approximately 25 to 30 percent over solid-planted corn.

Soybean plant characteristics were also not affected by the variety of corn. The pattern of intercropping played the major role. Different vegetative characters were affected by the intercropping pattern. Pod number and seed number per plant were about 30 to 50 percent more in solid soybeans than in intercropped culture. Intercropping decreased seed yield per plant more in late-maturing than in early-maturing cultivars. Seed yield of intercropped soybean reached one third to slightly more than half that of solid cropping. There was a slight effect of intercropping on oil content but protein content increased in intercropped soybean. Total yield of intercropped corn and soybeans was higher than both solid crops. This occurred in all intercropping patterns. Mean Land Equivalent Ratio varied from 122 to 131 percent in different seasons. The increase in total yield of intercropped species was due mainly to gain in yield of intercropped corn.

BREEDING SOYBEAN VARIETIES FOR VARIOUS CROPPING SYSTEMS IN SOUTHEAST ASIA

R. M. Lantican, Department of Agronomy and Institute of Plant Breeding, University of the Phillippines at Los Banos, College, Laguna, Phillippines

In Southeast Asia, soybeans are grown in various cropping systems: 1) mainly in rice paddies immediately following a crop of rice, often under conditions of zero or minimal tillage (seeds dibbled into the mud) and reliance on residual moisture; 2) as intercrops with other annuals like corn and sugarcane at partial sunlight; and 3) as monocrops. Soybeans are also being considered for planting in unoccupied spaces under coconut, oil palm and rubber trees in plantations.

Breeding and selection programs are usually undertaken under idealized, non-stress conditions of a monocrop system. It is doubtful if varieties evolved under such systems would fit equally well to the conditions of other cropping systems where stress factors such as shading, nutrient competition, poor soil tilth and insufficietnt moisture are crucial to varietal adaptation.

The first-stage selection program should exploit known varietal features that relate to general fitness and greater yield stability in performance over a range of cropping systems and seasonal patterns. The following are important selection indices and values should fall within the acceptable range specified: 1) maturity period from 80 to 95 days; 2) moderate vegetative development at leaf area index (LAI) values from 3 to 5; 3) high harvest index (HI) from 30 percent and higher; and 4) high seed weight from 150 gm per 1000 seeds and higher. Wide fluctuations in values beyond the above range can occur owing to seasonal influences, mostly due to rainfall and photoperiod differences. Values beyond these ranges especially late maturity of up to 130 days, high LAI at 8 and low HI will make for erratic behavior of varieties over changing environments.

Second stage evaluation of first-stage selections should be specific for each cropping situation in order to exploit further unique specialties for adaptiveness or tolerance to stress complexes such as shading and nutrient competition effects with intercropping and soil compaction and drought in the case of paddy field cultivation. Under shade conditions of intercropping, effects on seedweight and pod number are evident, with reductions of about 38 and 12 percent, respectively. Thus, these two characters may be used as indices in selection. On the other hand, under moisture stress conditions in the paddy, the important indicators for varietal adaptation are: 1) seed weight as positively influenced by a fast rate of seed filling, and 2) pod number. Varieties which deposit more than 3.5 mg of dry matter per day per seed and which attain maximum seed weight in only 40 days do well under paddy conditions. Obviously, a fast rate of seed filling confers an advantage in avoiding the critical period of moisture stress late in the growing season.

NODULATION, NITROGEN FIXATION AND YIELD OF SOYBEAN AS INFLUENCED BY DIFFERENT INOCULA APPLIED AT VARIOUS PLANTING DEPTHS

K. A. Ajam and S. M. Damirgi, Department of Soil Science, University of Baghdad, Iraq

High soil temperatures (35 C) at soybean planting time (May 15 to June 15) in central Iraq are one of the outstanding factors which limit the viability and the effectiveness of *Rhizobium japonicum* strains applied to soils as inoculum.

To evaluate and minimize the adverse soil temperature effect on soybean nodulation, N-fixation and yield, five different kinds of inocula were used under field conditions: 1) a liquid mixture of 1:1:1 ratio of *R. japonicum* serogroup 110, 123 and 216; and 2) the other four inocula were silt carriers mixed with 30 percent composed city refuse, 20 percent composted date palm leaflets, 30 percent composted alfalfa hay, and 100% silt, respectively. Four different methods of seed inoculation were used with each inoculum at planting date: 1) seed coating, 2) application of the inoculum on the seeds in the planting row, 3) application of inoculum to a 7-cm depth in the planting row, and 4) application of the inoculum to a 10-cm depth in the planting row.

After 57 days from planting nodule number and dry weight of nodules, dry matter and total nitrogen of plant tops and percent of *R. japonicum* serogroups initiated nodule formation were determined. After harvesting pod number, weight of 100 seeds and seed yield were determined. Results show that the higher numbers of nodules on soybean (var. Lee) were formed with all inocula when they were placed at the 10-cm depth, however, at this depth the highest nodule number was obtained with treatments receiving liquid inoculum and the least number with 100 percent silt carrier. Other variables under study were influenced differently with various treatments.

EVALUATION OF NARROW- AND BROAD-LEAFLET ISOLINES OF SOYBEANS [*GLYCINE MAX* (L.) MERRILL]

F. A. Mandl and G. R. Buss, Agronomy Department, Virginia Polytechnic Institute and State University, Blacksburg, Virginia 24061

The objective of this study was to evaluate the effect of the leaflet type on seed yield and other agronomic traits. In all work reported previously using isolines to study the effect of the narrow-leaflet type, the narrow-leaflet lines were developed through back-crossing. In this study, crosses were made between five broad-leaflet genotypes adapted to Virginia and a narrow-leaflet line of Mississippi (D64-4731). Narrow- and broad-leaflet plants were selected within the progeny of heterozygous broad-leaflet F_4 and F_5 single plants, respectively. In the next generation, a narrow-leaflet progeny row and a non-segregating broad one, which traced to the same F_4 and F_5 single plant, were selected and considered a pair of isolines. Therefore, the genetic background of each pair is a random recombination of the narrow- and broad–leaflet genotypes of the parents.

Ten pairs of isolines were evaluated in field plots in 1976. Six additional pairs were included in 1977 and 1978. A split-plot design with pairs as main plot and leaflet type as sub-plot was used. Data were recorded for seed yield, seed size, plant height, maturity date, and lodging. Data on yield components, leaf area, oil and protein content were obtained from six pairs of one cross in 1978. The 1978 data are not available at this writing, but no significant differences in seed yield between leaflet types were obtained in the first two years. The narrow-leaflet lines yielded 26.6 kg/ha more and 12.6 kg/ha less than the broad-leaflet lines in 1976 and 1977, respectively. Narrow-leaflet isolines were significantly smaller in seed size (1.3 and 0.9 g/100 seeds) and shorter in height (1.5 and 2.3 cm) in 1976 and 1977. In 1976, the narrow-leaflet lines averaged significantly earlier in maturity (1.3 days), but in 1977, there was no difference between the two types in maturity. The average difference in lodging score was very small both years, but the broad-leaflet types lodged significantly more in 1977.

The results obtained in the first two years indicate no difference between the leaflet types in terms of seed yield. The differences in the other traits, in spite of being statistically significant in several cases, do not appear to indicate any practical advantage of one leaflet type over the other.

SEED DEVELOPMENT AND GERMINATION OF SOYBEANS AT VARIOUS FILLING STAGES

K. Y. Park, Crop Experiment Station, Suweon 170, Korea

Six soybean cultivars were used to study their seed development and germinability after flowering in 1977 and 1978. The earlier varieties showed the faster seed development, but indeterminate types resulted in slower development. The earliest germinability was found in 15 to 25 days after flowering when dried with the pod shell. The germination of the seeds dried without pod shell, however, was inhibited significantly in early seed filling stages. Regular germinability of the Hill variety in a cold test was 40 days after flowering, or 10 to 15 days later than the first germinability found.

VARIETAL DIFFERENCES FOR GERMINATION AND EMERGENCE IN SOYBEAN [*GLYCINE MAX*] UNDER FIELD, GREENHOUSE AND LABORATORY CONDITIONS

R. N. Trikha and S. Shanmugasundram, Soya Production and Research Association, Bareilly, U.P. India and The Asian Vegetable Research and Development Center, Shanhua, Tainan, Taiwan, R.O.C.

Varietal differences for germination and emergence were studied among ten soybean [*Glycine max*] varieties harvested in different seasons and kept in storage for 15, 11 and 9 months. Seeds were treated with and without fungicide and sown at The Asian Vegetable Research and Development Center in autumn 1977-78 under greenhouse, field and laboratory conditions. Germination was highest in the laboratory followed by greenhouse and field. The correlation was highly significant. Seed treatment with fungicides had improved percentage of germination with emergence. The treatment minimizes the number of fungal infected seeds but increases the number of bacterial infected seeds. The season of seed production and the duration of storage are the two vital factors determining the percent emergence. A cool, dry season of seed production with a shorter storage period was the best for higher emergence. Increase in seed size affected germination adversely. The initial seed vigor in the form of hypocotyl length was inhibited by the fungicide treatment.

Genetic differences are probably present and the inherent seed viability is an essential factor in determining the plant stand. It may be worthwhile to identify these traits for incorporation into the new varieties. The seeds of some varieties appear to be less susceptible to fungal and bacterial infection than others. Adequate plant stand is essential for high yield. Under the field conditions, besides the seed quality, there are factors such as moisture, temperature, soil-borne organisms, etc., which play an equally important role in the percent emergence.

IDENTIFICATION OF SOYBEAN GENOTYPES WITH SUPERIOR SEED QUALITY

E. H. Paschal, II and M. A. Ellis, Texas A&M University Agricultural Research and Extension Center, Beaumont, Texas 77706 and Department of Crop Protection, University of Puerto Rico, Mayaguez, Puerto Rico 00708

Production of high quality soybean [*Glycine max* (L.) Merr.] seed is difficult in humid tropical and subtropical environments. High temperature and relative humidity during the final stages of growth are not conducive to the production of high quality seed. These conditions are, however, conducive to the development of seedborne diseases which reduce soybean seed quality. We postulated that since the soybean has been cultivated for ages in the Asian tropics and subtropics, genotypes should have evolved which possess desirable seed quality attributes. The objective of this research was to determine the extent of genetic variation in the incidence of soybean seed infection by fungi and loss of seed viability under tropical conditions.

A total of 396 accessions from maturity groups VIII, IX, and X of the USDA soybean germplasm collection were grown at Isabela, Puerto Rico (18° N) to produce seed for a preliminary screening. Seed of each accession were surface sterilized and placed on potato dextrose agar (PDA) in petri dishes. After incubation at 25 C for 7 days, data were recorded on in vitro germination percentage and the incidence of internally seedborne microorganisms. Twenty-one accessions with superior seed quality and three checks were selected for further testing in replicated trials planted in February and July 1976. Treatments were included in which harvest was delayed until 2 or 4 weeks after maturity. Data were recorded on in vitro germination and the incidence of internally seedborne microorganisms as described for the preliminary bioassay. In addition, sand emergence, field emergence, and agronomic characteristics of the lines were also recorded.

The average incidence of internally seedborne fungi of the 24 accessions increased from 9 percent in the seed harvested at maturity, to 29 percent in the seed from the 2-week delayed harvest, and 45 percent in the seed from the 4-week delayed harvest. The average incidence of *Phomopsis* sp., the causal organism of pod and stem blight of soybeans, was 2, 9, and 20 percent in the nondelayed, 2 and 4-week delayed harvest, respectively. Pod and stem blight is considered the principal disease affecting soybean seed quality. Average field emergence of the 24 entries was reduced 14 and 37 percent by the 2 and 4-week delays, respectively. Substantial genetic variation was encountered for the seed quality characters measured. Sand emergence of four of the accessions—PI 205908, PI 205912, PI 219653, and PI 239235—was not reduced significantly when harvest was delayed 4 weeks, but that of the three checks was reduced 48 percent. Field emergence percentage was correlated negatively with the incidence of fungi, the incidence of *Phomopsis* sp., seed size, and seed quality score.

EFFECT OF FREQUENCY AND AMOUNT OF IRRIGATION ON SOYBEAN

B. Yazdi-Samadi and K. Saadati, College of Agriculture, University of Tehran, Karaj, Iran

One of the limiting factors for increasing acreage and yield of crop plants in many parts of the world is water shortage. Water stress at different stages of growth produces different results. Many investigators observed that soybean yield increased significantly when irrigation was high during the blooming stage, and increased irrigation at the pod-filling stage had more effect on yield than any other stage of development. The objective of this work was to determine the effect of number and amount of irrigation at different growth stages on yield and other agronomic traits of soybean.

Two experiments were conducted to study the effects of frequency and amount of irrigation in different growth stages in soybean [*Glycine max* (L.) Merr]. In Experiment I (in 1974) five irrigation treatments consisting of 4, 6, 8, 11 and 15 irrigations and in Experiment II (in 1975) seven irrigation treatments: 4, 5, 6, 7, 8, 11 and 14 irrigations (with 179, 293, 396, 491, 573, 738 and 839 mm of water, respectively) were used. Several agronomic traits including number of pods per plant, nodule number of roots, plant height, number of seeds per pod, seed weight (g/seed), seed yield (kg/ha), and oil percentage were studied. Results of the experiments showed that: 1) Higher amounts of irrigation increased nodule count per plant, plant height, number of pods per plant, seed weight and seed yield; however, it had not affected the number of seeds per pod and oil percentage very much. 2) Yield increase in treatments with higher amounts of water was due mostly to the increase in the number of pods per plant and seed weight. 3) In Experiment I, one irrigation before flowering produced a yield of 1248 kg/ha, and extra irrigations before and after flowering increased yields up to 4200 kg/ha. The same trend of yield increase was also observed in Experiment II. 4) it was noticed in both experiments that irrigation at the vegetative stage of the crop was important in increasing yield and may not be neglected, and 5) The comparison of different irrigation treatments in Experiment II proved that irrigation at the end of flowering which corresponded to the pod-filling stage was more important than at other stages in increasing soybean seed yield.

SOYBEAN RESPONSE TO WATER DEFICITS DURING SPECIFIC GROWTH STAGES

P. L. Sutherland and R. E. Danielson, Department of Agronomy, Colorado State University, Fort Collins, Colorado

Until recent years, soybean production has been limited to the midwestern and southern states. There exists a great potential for expanded production of soybeans into the irrigated areas of the western United States because of the demand for soybeans as a source for protein and resulting expansion of markets. With this existing potential for expanded soybean production, one of the most important factors for attaining high yields in a semi-arid climate is proper water management. However, it should be recognized that much of the research pertaining to water relations and water management has been conducted in the humid, high rainfall locales.

Field studies were established during a five year period at Fort Collins, Colorado, to evaluate the influence of irrigation timing on the yield components, yield, and quality in soybeans [*Glycine max* (L.) Merr.] under semi-arid conditions. In 1974, five irrigation treatments and three within-row plant populations were combined into a randomized block, split-plot field experiment. The levels of irrigation included a non-irrigated treatment, a well-irrigated treatment and treatments where water was withheld prior to flowering, during flowering, and during pod and bean development. In 1975 thirteen irrigation treatments and two within-row plant populations were combined into a randomized, split-plot design providing for water deficits during the entire season, during three stages of growth, during two stages, during one stage of growth, and a well-irrigated treatment. In 1976, 1977, and 1978 a "Line Source" continous variable experimental design was used to evaluate water deficits during specific stages of growth as well as amounts of water applied during irrigation. A single sprinkler line extended through the center of the field parallel to the rows. The experimental design provided for uniform water application parallel to the sprinkler line with decreasing application with increasing distance from the line. Thus, the outside rows received little or no water during an irrigation, while rows immediately adjacent to the sprinkler line received high levels of water. Seven irrigation treatments were incorporated over the three growing seasons, allowing for periods of water deficits during critical stages of growth. Soil water conditions were monitored during the growing season using neutron attenuation during all years. Bean yields were obtained at maturity with samples taken for protein and oil analysis. In addition, samples were obtained for evaluation of the yield components.

From the results of the field studies several conclusions can be made. First, the pod and bean development stages of growth are the critical growth stages that require the maintenance of an adequate soil water supply. Secondly, a water deficit during flowering increases bean yield, while water applied during flowering depresses yield. Thirdly, high soil water content throughout the pod and bean development stages delays maturation which could decrease yields if the growing season is restricted.

SOYBEAN RESIDUES AS A SOURCE OF MOST OF THE NITROGEN NEEDED FOR CORN IN THE STATE OF SAO PAULO, BRAZIL

H. A. A. Mascarenhas, R. Hiroce, N. R. Braga, M. A. C. de Miranda, C. V. Pommer, and E. Sawazaki, Instituto Agronomico, Campinas, Sao Paulo, Brazil

Soybean utilizes atmospheric nitrogen in a symbiotic relationship with *Rhizobium japonicum*. There is considerable interest in Brazil in determining the benefit to non-legumes grown following soybeans, especially so because of the high cost of nitrogen fertilizer. To verify the contribution of crop residues including roots and nodules to N nutrition of a non-legume subsequent crop, corn was planted after one to four years of continuous soybean cultivation at two localities. At planting time at one locality in each of the four experiments, 10 kg/ha of N was applied whereas at the other locality 16 kg/ha of N was used as a starter together with the PK fertilization. The N levels utilized as side dressing were 0, 20, 40, 60 and 80 kg/ha applied 30 days after germination in the form of ammonium sulphate.

The results showed that at one locality corn yields after one year of soybean cultivation presented a tendency for response to nitrogen applied as side dressing. After two or more years of soybean cultivation, there was no significant nitrogen response in crop production. At the other locality nitrogen side dressing did not promote significant increases in yield even after one year of soybean cultivation. A combined analysis of production for corn at the two localities showed small increases in yield due to nitrogen application as side dressing which was neither significant nor economical. There was an increase in the yield of corn where soybeans had been planted previously and this increase was directly proportional to the number of years that soybeans were cultivated.

PRODUCING SOYBEANS IN CALIFORNIA

W. D. McClellan, University of California, Cooperative Extension, Visalia, CA

The commercial acreage of soybeans in California has increased from five acres in 1973 to over 10,000 acres in 1977. The majority of these soybeans were planted from mid-June to mid-July as a double crop following the harvest of winter cereals. The increased value of soybeans in recent years and the emphasis on using them in a double crop situation may make the crop a successful addition to field crop rotation patterns in California. While the potential of this crop is being studied in many areas of the state, the information presented here will deal with soybeans grown on the east side of the San Joaquin Valley.

Williams and Amsoy 71 have been the two varieties planted most frequently in commercial fields. These varieties reach harvest maturity 120 to 130 days after planting and have yielded well both in research plots and in growers' fields. Studies on row spacing and plant population with the variety Williams have shown yield potential over 4,000 pounds per acre (66 Bu/A) on 12" rows and a population of 140,000 plants per acre when grown as a double crop after barley. The highest yield obtained in a commercial planting was 3,657 pounds per acre (61 Bu/A). This field of Williams was planted with a grain drill (7" spacing) following the harvest of wheat. Low yields were obtained in fields with 38 or 30" row spacing and low plant populations (less than 70,000 plants/A). The growers who obtained the highest yields (with either Amsoy 71 or Williams) had closer row spacing and higher plant populations. In addition to having excellent plant density, these growers had few weeds and did not stress the plants for water during the season.

In 1975-76, harvest losses assessed in several commercial fields ranged from 62 to 580 pounds per acre. Harvest losses were primarily due to pods set too low to the ground and uneven ground (especially in bed plantings). Shatter, often considered a major limiting factor in soybean production in California, is minimal with the varieties Amsoy 71 and Williams when planted mid-June to mid-July. Harvest losses were minimized where close row spacing, high plant population and flat plantings were used. This resulted in the fewest pods near the ground and allowed the cutter bar to be set close to the ground. In 1975 harvest losses averaged 397 pounds per acre (17 percent of the total yield) where the soybeans were furrow irrigated; this compared to 251 pounds per acre (9.7 percent of the total yield) where the crop was planted on the flat and flood irrigated.

The marketing of soybeans grown in California has developed rapidly along with the increase in acreage. Several commodity brokers have indicated an interest in soybeans in California. The majority of the acreage (over 10,000 acres) grown in 1977 was contracted by the grower prior to planting. Most of the contracts were on an acreage basis, not actual tonnage. Growers received from $200-250 per ton in 1976-77 seasons. The attractive price and the contract on an acreage basis provided encouragement to new growers in this trying crop. A cost study survey conducted in 1976 showed the growers' cash cost for soybeans planted as a double crop ranged from $80.04 to $162.25 per acre. Several growers also included the cash costs for two other double crops grown in the San Joaquin Valley, corn silage and milo. In each instance, the cash costs for these crops were $25-30 per acre *higher* than soybeans. The additional costs were due to: 1) higher water usage, 2) use of nitrogen fertilizer, and/or 3) drying costs (milo) at harvest.

Spider mites have historically been considered the major pest problem of soybeans in California. Grower fields in 1975-76 were surveyed for mites and other insects on a weekly basis. In 1975, most growers controlled mite populations using a preventative approach—applying a miticide before economically damaging levels were reached. In 1976 and 1977 growers began to utilize chemical control methods only if the situation warranted it based on the weekly field evaluations. Of the surveyed fields, less than 25 percent were treated with a miticide. This successful use of field survey techniques to determine the extent of mite and other insect infestations was an effective method of minimizing the need for chemical pest control measures. In summary, spide mites have not been a limiting factor in the growing of soybeans as a double crop on the east side of the San Joaquin Valley. With few exceptions other insect pests did not reach damaging levels in the fields surveyed. Registration of miticides for use on soybeans is still a problem, but efforts are being made toward obtaining registration.

There is a definite need for new varieties in California that have some or all of the following characteristics: 1) higher yield potential than Amsoy 71 and Williams; 2) maturity reached in 100-110 days rather than the 125-130 days for Williams and Amsoy 71; 3) lodging resistance; 4) good seed quality; 5) bottom pods set high (6 to 8 inches) off the ground; 6) spider mite resistance; 7) shatter resistance; 8) salt tolerance; and 9) a determinate type of growth. There are several other characteristics that could be listed in addition to these.

PROTEIN AND OIL

Invited Papers

WORLD TRADE IN SOYBEANS

R. E. Bell, Riceland Foods, Stuttgart, Arkansas

Soybeans is one of the fastest growing items in world trade. World trade in soybeans expanded at an average rate of 11 percent per annum during the past five years. World trade in soybeans and soybean products totaled 35.6 million metric tons during 1977/78. Its value was approximately $8.5 billion.

As a result of the rapid growth in world trade in soybeans and soybean products, world consumption of soybeans increased 45 percent during the past five years. World soybean consumption was a record 77 million metric tons in 1977/78, with 85 percent of the beans being crushed into oil and meal. Only in East Asia is there a high usage of whole beans for direct food consumption. Elsewhere, the meal produced from soybean crushings is used primarily as high-protein feed for livestock, while the oil is used for direct human consumption.

The recent growth in world trade in soybeans would not have been possible without a sharp expansion of soybean production in Brazil. Production in Brazil increased thirteen-fold during the past decade and should reach 13.5 million metric tons in 1979. Nearly 80 percent of the increase in Brazilian production has gone into export either as beans, oil or meal, with most of the exports going to Europe. The United States also expanded its production and exports of soybeans and soybean products during the same period, but not to the extent of Brazil. Production in the United States in 1978 was a record 49.3 million metric tons, and U.S. exports of beans, oil and meal will total nearly 27 million metric tons during 1978/79.

The growth in world trade in soybeans caused soybean prices to more than double in nominal terms during the past five years. The increase in real terms, however, was only about 45 percent. The higher prices encouraged an expansion of soybean acreage in both the United States and South America. World soybean acreage increased nearly 25 million acres during the past five years. If recent trends continue, soybeans will soon replace corn in the United States as the leading crop in terms of harvested acres. It has now replaced both wheat and corn as the leading cash crop in the United States.

World consumption of soybeans expanded at an average rate of 8 percent per annum during the past five years. This demand has been met primarily through an expansion in soybean acreage. If this consumption trend is to continue in future years, it is imperative that a breakthrough be achieved in soybean yields per acre.

SOY PROTEIN AND OIL: THE PROCESSORS' VIEW

W. H. Martinez, SEA/AR, USDA, National Program Staff, Beltsville, Maryland

Protein and oil are the primary products of the soybean processing industry. Historically, the protein product has represented the major portion of the value of the seed although the unit value of the oil is greater. Seed quality, which is both genetically and environmentally controlled, is the major determinant of product yield and product quality. The relative proportion of seed components—protein, oil carbohydrate and fiber—is critical to yield and the production of products with desired protein levels. Oil content, therefore, must be consistent with levels that are acceptable to the economics of processing, and a minimal level of carbohydrate is desirable.

Composition of the components is also critical to the acceptability and value of the product. The major factor limiting the stability, versatility, functionality and cost of processing soybean oil is the linolenic acid content. Two variables—world supply of fats and oils, and cost and nutritional acceptability of the hydrogenated products—are additional factors in determining the value of all vegetable oils, but the phospholipid content and composition of extracted oil is particularly important to the quality of refined soybean oil.

Sulfur amino acid and trypsin inhibitor contents are major quality factors in soybean protein products when used as feed or food. Other factors such as the concentration of complex oligosaccharides and constituents that contribute to undesirable flavor are important in the developing soybean food protein market. The need to optimize the quality of soybean protein products with respect to each of these factors affects processing and energy requirements and yield.

Ideally, the processor desires a soybean with increased protein and increased oil obtained at the expense of carbohydrate and fiber (hull), and free of all deleterious constituents. The compatability of this ideal with genetic capability, planting seed quality, and seedling stand has yet to be resolved.

SOYBEAN PROTEINS IN HUMAN NUTRITION

W. J. Wolf, Northern Regional Research Center, SEA/AR, USDA, Peoria, Illinois 61604

Soybean proteins play an important, although indirect, role in the U.S. diet because soybean meal is a major source of the protein in feed that is converted into animal proteins such as meat, milk and eggs. However, as prices continue to rise, the animal is being by-passed and soybean proteins are used directly to extend or replace traditional animal proteins. Because animal proteins are often used as standards of nutritional quality, the question of nutritional equivalency is being asked of soybean proteins. It is widely accepted that soybeans are limiting in methionine and cystine based on rat bioassays, but man has a lower requirements for the sulfur amino acids than the rat. Indeed, recent studies with humans indicate that soybean proteins are of higher nutritional value for man than previously believed.

Another nutritional concern in direct consumption of soybean proteins by humans is the residual trypsin inhibitor activity of processed soybean products. In several species, the pancreas is the primary organ affected by high levels of trypsin inhibitors such as occurs on ingestion of raw soybean meal. Possible effects of trypsin inhibitors in humans are unknown, and research is needed to determine whether the low levels of inhibitor activity in processed soybean foods are of any consequence.

PRICING SOYBEANS ON THE BASIS OF CHEMICAL CONSTITUENTS

T. E. Nichols, Jr., North Carolina State University, Raleigh, North Carolina 27650

Growers of high-quality soybeans may one day be in a position to receive price premiums because of the widespread use of a quick test of protein and oil levels. Similarly, farmers with soybeans having low oil and protein content will receive a discount for their beans. This should minimize some of the pricing inequities and problems for both growers and processors in marketing their products.

The value of a bushel of soybeans depends on how much oil and protein are in the beans and the expected market price of soybean oil and meal. Yet, oil and protein levels currently are not factors in the quality determination and pricing system of soybeans. Large quantities of soybeans and other oilseeds change hands daily with buyers and sellers alike lacking precise information about the value of the chemical constituents. Until recently traders were unable to make these determinations due to the difficulties of testing for oil and protein levels at the time of delivery.

Instruments for quick and reliable measurements of moisture, protein and oil levels in oilseeds and grain are now available at reasonable costs. Their use will allow buyers to quickly determine quality differences among lots and adjust the price paid for soybeans based on constituent values.

The objectives of this paper are to: 1) examine the joint product relationship of soybean meal and oil; 2) analyze the effects of using a system of soybean discounts and premiums on producers, handlers, and processors; and 3) estimate the cost and benefits to society of adopting a component pricing system for soybeans.

BIOLOGICAL REALITIES OF MEETING MARKET AND WORLD DEMANDS FOR SOY PRODUCTS

C. A. Brim, SEA/AR, USDA, Southern Region, North Carolina State University, Raleigh, North Carolina 27650

Cultivar development is an evolving process that is responsive to changes in production problems and in demand for the final products, protein and oil. Correlated variation between traits of economic importance has an important impact on progress in developing cultivars that meet these demands. The correlated reponse between traits may be due to similar genetic causes or to similar responses to environmental influences. Negative correlations, even though they may be small, that are the result of genetic causes are of particular concern to the plant breeder.

This paper will review the constraints to selection progress imposed by correlated responses between chemical components of the seed and between these components and yield. Constraints to progress imposed by additional selection criteria relating to quality factors and to solutions of production problems will also be reviewed.

PLANT PATHOLOGY

Invited Papers

THE ROLE OF *PHOMOPSIS* SP. IN THE SEED ROT PROBLEM

A. F. Schmitthenner, Department of Plant Pathology, Ohio Agricultural Research and Development Center and The Ohio State University, Wooster, Ohio 44691

Poor seed quality in the northern U.S.A. is primarily the result of infection with an undescribed species of *Phomopsis. Diaporthe phaseolorum* var. *sojae* (Dps) and *D. phaseolorum* var. *caulivora* (Dpc) are minor components of the seed rot complex (less than 20 percent of isolates obtained). These fungi can mold seed while still in the pod, but cause the greatest damage through latent seed infection which developes into seedling rot during germination. Germination of seed is inversely proportional to the percentage infection with *Phomopsis* and *Diaporthe* where this seed quality problem exits.

Phomopsis differs from Dps and Dpc by the absence of the perfect stage in both nature and culture, by pycnidial characteristics, and by stroma development and other cultural characteristics. *Phomopsis* survives from crop to crop in the pycnidial stage in soybean straw. Spores from these pycnidia cause latent symptomless infection on stems and petioles of the growing crop. Pycnidia form on fallen petioles and produce fresh inocula throughout the growing season. Both primary and secondary inocula infect young pods as they develop, but infections remain latent until pods begin to mature. If wet, humid conditions occur at this time colonization of pods and infection of seed occurs. Dps and Dpc are less of a problem because they persist from crop to crop in the perithecia stage, which is not abundantly produced, in soybean straw and produce no secondary inoculum during the growing season.

Infection of soybean seed with *Phomopsis* and *Diaporthe* can be reduced by crop rotation and plowing down residue to decrease inoculum; planting late varieties or planting late so that the crop matures late in the season when less seed infection occurs; by harvesting the crop as soon as it matures; by providing adequate soil potash; and by protecting the crop with fungicides. Germination of *Phomopsis* infected seed can be improved by seed treatment.

SOYBEAN DISEASES IN BRAZIL: STATUS, DISTRIBUTION AND RESEARCH PROGRAM

C. C. Machado, A. M. R. Almeida, L. P. Ferreira and J. Yamashita, EMBRAPA, Centro Nacional de Pesquisa de Soja, P.O. Box 1061, 86.100, Londrina, Parana, Brazil

Among the many problems affecting soybean yield in Brazil, diseases are undoubtedly a very important factor. The most important soybean diseases in this country are the following. Bacterial diseases: 1) Bacterial Blight (*Pseudomonas glycinea*), is widespread in all production areas, and is considered to occur with the greatest frequency; 2) Bacterial Pustule (*Xanthomonas glycines*) is also widespread but with lower frequency, but does occur with higher intensity in the "Cerrado" region of Central Brazil. Fungal diseases: 1) Dead Patch (*Rhizoctonia solani, Fusarium* spp) occurs mainly in the State of Rio Grande do Sul, where losses may reach up to 40 percent, and also in south Parana and Mato Grosso do Sul; 2) Brown Spot (*Septoria glycines*) is widespread in all production areas, increasing its frequency and severity from year to year; 3) Frogeye (*Cerospora sojina*) is widespread but with higher frequency and severity in the States of Parana and Mato Grosso do Sul; 4) Pod and Stem Blight (*Phomopsis sojae*) is widespread with high frequency and severity within the country; 5) Whetzelinia Stem Rot (*Whetzelinia sclerotiorum*) occurs sporadically but in south Parana in 1977 and 1978 it caused losses up to 50 percent; 6) Charcoal Rot (*Macrophomina phaseolina*) is widespread within all production areas with higher frequency and severity in north Parana and central-west regions; 7) Anthracnose (*Colletotricum dematium* var. *truncata*) is widespread but with low frequency and severity in all areas. Virus diseases: 1) Soybean Mosaic is widespread in all production areas; 2) Braziliam Bud Blight is also widespread with variable frequency and severity from year to year and from place to place, sometimes causing severe losses. Nematode Diseases: *Meloidogyne* spp. are the most important and widespread genus of nematodes within the country. *M. javanica* and *M. incognita* are the prevalent species, occurring with higher frequency and severity in Central Brazil. Nine other nematode genera associated with soybeans have been described in Brazil. Several other bacterial fungal, virual, and nematode diseases occur in the various soybean production areas, but are considered of minor importance. The Research Program on Soybean Pathology presently in development in Brazil by the National Soybean Research Center is based mainly on: 1) diseases survey; 2) research on sources of resistand; 3) epidemiology and control; and 4) seed pathology.

RESEARCH ON *PHYTOPHTHORA* ROOT AND STEM ROT

B. L. Keeling, Delta Branch Experiment Station, Stoneville, Mississippi

This presentation includes a discussion of the following: a) techniques used to isolate and classify pathogenic races of *Phytophthora megasperma* var. *sojae;* b) the race situation as it now exists; c) testing for varietal resistance to a single race, a combination of races simultaneously, and for lower levels of resistance; and d) a report on new virulences isolated from soybeans in Mississippi.

Contributed Papers

SEED AND SEEDLING DISEASES—A POTENTIAL THREAT IN SOYBEAN PRODUCTION IN WARM HUMID AREAS

M. N. Khare, Department of Plant Pathology, Jawaharlal Nehru Agriculture University, Jabalpur, M.P. India

Soybean has attracted the attention of cultivators in India. The State of Madhya Pradesh has proved best for soybean production and is occupying the maximum acreage in the country.

The seeds produced in various parts of the state were examined for the associated microflora and quality. So far 32 fungi and three bacteria have been observed associated with seeds of which *Macrophomina phaseolina, Sclerotium rolfsii, Myrothecium roridum, Phomopsis sojae, Cercospora* spp., *Colletotrichum dematium,* f. sp. *truncata, Cornespora cassicola, Phoma* sp., *Fusarium moniliforme, F. semitectum, F. equilseti, F. oxysporum, Xanthomonas phaseoli, Pseudomonas* spp. were important. Percentage association of important pathogenic fungi was determined and their relationship with seedling damage was established. Several species of *Aspergillus* and *Penicillium* were also associated as storage fungi.

The seed-borne fungi and bacteria resulted in seed rot, seedling blight, hypocotyl rot, foot rot, root rot, stem rot, pre- and post-emergence losses of various types and diseases at later stages of crop growth.

High temperature and high humidity at the planting time proved detrimental for seed and seedling rots, Rains towards maturity of the crop were responsible for pod diseases leading to seed damage. Seed produced in humid areas had poor storability due to the associated microflora.

TRANSMISSION OF *PSEUDOMONAS GLYCINEA* VIA SOYBEAN SEEDS

B. W. Kennedy, Department of Plant Pathology, University of Minnesota, St. Paul, Minnesota 55108

Although seed transmission of bacterial blight of soybean *(Pseudomonas glycinea* Coerper) has long been known, the nature of symptoms on infected seed and the efficiency with which individual seeds with specific symptoms will transmit bacterial to the seedling has not been clearly shown. The objective of this investigation was to elucidate such symptomatology and transmission.

Evaluations were made by use of a variety of seedlots, determining symptomatology of infection via artificial inoculation of developing pods and by bioassaying an array of "blemishes" from naturally infected seeds. All seeds showing abnormabilities were divided into categories characterized as moldy, brown discolored, green shriveled, yellow shriveled erumpent pustules, and miscellaneous. Also, any brownish-yellow "grease spots" noted alone or in combination with these categories were designated separately. Individual seeds were surface sterilized by being dipped in 70 percent alcohol, then placed in 0.5 percent sodium hypochlorite for one minute, dried on sterile paper towels, and then placed individually into two ml nutrient broth overnight at 27 C. Broth was then injected into leaves of Acme soybean seedlings in the greenhouse. Seedlings for this assay were from seeds produced for three generations in the arid western USA and were free of bacterial pathogens. For transmission studies, blemished seeds in the various categories were grown in sand in the greenhouse and observed for blight symptoms.

A summary of total seed tested by broth method is as follows: "grease spot"—436 seeds tested, 231 infected; "blemish"—376 seeds tested, 130 infected; control (no symptoms)—289 seeds tested, 10 infected. A summary of seeds tested by germination of seeds and observed for infection in the seedlings is as follows: "grease spot"—84 of 165 seeds germinated and 59 of those germinating were infected; "blemish"—42 of 85 seeds germinated and 8 of those germinating were infected; control—232 of 250 germinated and none of those germinating were infected.

Health appearing soybean seeds may carry *P. glycinea* internally but such seeds infrequently give rise to a diseased seedling. On the other hand, seeds with any of a variety of abnormalities, especially those possessing "grease spot," may be suspected of harboring the bacterium and transmitting the disease.

BENZENE AND ETHANOL AS FUNGICIDE CARRIERS IN DORMANT SOYBEAN SEED

James J. Muchovej and Onkar D. Dhingra, Departamento de Fitopatologia, Universidade Federal de Vicosa - 36570 Vicosa, MG - Brasil

Soybean cultivar 'UFV-4' with unbroken seed were soaked in benomyl, thiabendazol, chlorothalonil, PCNB or RH-2161 mixture with benzene or ethanol. The seeds were removed from the mixture after 0.5, 1.0, 1.5, 3.0, 6.0 or 24.0 hours. After evaporation of the solvents, seeds were washed with acetone to remove the fungicide from the seed surface. The seeds were cut laterally into half. The seed coat was removed from one half of the seed but not from the other half. Both seed halves were then placed on potato-dextrose agar seeded with *Trichoderma* sp., and incubated at 25 C. Zones of growth inhibition of the fungus around the seed halves were measured after 72 hours.

A soybean seed lot with a 15 percent in vitro germination with 45 percent of the seed infected with *Phomopsis sojae* and 29 percent with *Fusarium semitectum* and germination potential of 66 percent were soaked in each solvent fungicide mixture for one hour. After treatment the seeds were divided into two portions. One portion was washed with acetone removing fungicide from seed surface, whereas the other portion was not washed. They were planted in sand and percentage emergence was determined 10 days after planting.

Soaking soybean seed in benzene or ethanol for up to 24 hours did not reduce germinability of seed. Activity of benomyl, thiabendazol, and RH-2161 was found in the seed halves with the seed coat attached, but not around seed halves without a seed coat. This suggests that the fungicides accumulated only in the seed coat of soybean seed, and did not reach the surface of the embryo. The percentage emergence of seedling treated with benomyl, thiabendazol or RH-2161 in benzene or ethanol, or soaked in solvents alone was significantly higher than untreated seeds and similar to the germination potential. There was no significant difference in the percentage emergence whether the fungicide-solvent mixture was or was not washed off of the seed surface, indicating that the amount of fungicide accumulated in the seed coat was sufficient to control seed-borne fungi.

THE EFFECTS OF FUNGICIDE APPLICATION AND HARVESTING DATE ON YIELD, SEED QUALITY, GERMINATION AND EMERGENCE IN SOYBEANS

G. F. Nsowah, University of Science and Technology, Kumasi, Ghana

Seeds of Hardee and Bossier were sown in a randomized block design in the field on April 15, and those of Improved Pellican and TGM 627 on April 8 and 23, 1976, respectively, so that they would flower at about the same time. The plants were then subjected to the following treatments: 1) control where no fungicide treatment and no inoculation of *Cercospora kikuchii* were applied; 2) soybeans were inoculated with spores of *C. kikuchii* at flowering; 3) soybeans were sprayed with 0.2 percent benomyl at early and mid-pod stages of development; and 4) soybeans were sprayed with 0.2 percent benomyl at early-pod, mid-pod, late-pod stages and at two weeks before the pods were mature.

At maturity the pods were harvested and dried in an oven for 24 hours at 50 C. Seeds were then hand extracted and percentage smooth clean, purple stained, total stained, cracked, shrivelled, weathered and insect-damaged seeds was determined. Germination tests were carried out in the laboratory and seedling emergence tests were conducted in the greenhouse and in the field immediately after harvest and after three to twelve months under cold storage. The 0.2 percent benomyl application reduced the percent purple stained seeds but not the percent cracked seeds. Delayed harvesting did not significantly affect yield but reduced percentage seedling emergence. No visible symptoms of inoculation with *Cercospora kikuchii* were observed on the plants in the field.

EFFECT OF MATERNAL CYTOPLASM ON THE INHERITANCE OF RESISTANCE OF SCARLET RUNNER BEANS, *PHASEOLUS COCCINEUS* SUBSP. *COCCINEUS,* TO SOYBEAN RUST, *PHAKOPSORA PACHYRHIZI,* IN PUERTO RICO

Nader G. Vakili, USDA, Mayaguez Institute of Tropical Agriculture, Mayaguez, Puerto Rico 00708

Scarlet runner beans (SRB) *(Phaseolus coccineus* subsp. *coccineus)* are a cross-pollinated legume consisting of a highly heterozygous population. Soybean rust (SBR) *(Phakopsora pachyrhizi)* is endemic in the highlands of Puerto Rico. Severe field incidence of SBR in an SRB plot in 1976 resulted in 29 resistant plants out of 750. The frequencies of the responses of the open-pollinated seedlings was affected, with a highly significant x^2 value, by the responses of their maternal parents. Also, responses of the materal plants affected, with a highly significant x^2 value, the frequency of the occurrence of mature plant susceptibility among the seedlings selected for their resistant response to SBR. The F_1 progenies from reciprocal crosses had similar frequencies of response as the S_1 seedlings and again the effect of material response on seedling frequencies for SBR response was highly significant. These data indicate that response of SRB to SBR is governed primarily by plasmagenes. The presence of plasmagenes in a SRB plant conditions susceptible response to soybean rust. The segregation of S_1 seedlings from immune plants for susceptibility suggests that SBR-SRB interaction is also conditioned by nuclear-genes.

ENVIRONMENTAL FACTORS AFFECTING DISTRIBUTION OF SOYBEAN RUST, *PHAKOPSORA PACHYRHIZI,* ON HYACINTH BEANS, *DOLICHOS LABLAB,* IN PUERTO RICO

Nader G. Vakili, Research Plant Pathologist, USDA, Mazaguez, Puerto Rico 00708

Soybean rust (SBR) *(Phakopsora pachyrhizi)* is endemic in the highlands of Puerto Rico. A number of perennial and annual wild and edible legume species are affected by this rust. Hyacinth beans (HB) *(Dolichos lablab)* is the cultivated reservoir of SBR on the island. Soybean rust is most prevalent on HB at elevations of 500 to 750 m where average temperatures range between 19 to 22 C. Controlled climate chamber studies indicate that the HB-SBR interaction is affected by temperature. At 25 to 30 C, inoculated HB plants are mostly free of rust, while at 17 to 20 C they are severely infected. Soybean rust spreads during the rainy season among the HB population of the island. It is present all year in areas where average precipitation is 200 cm or more. During the rainy season, May to November, the rust may frequently be encountered at any elevation where average rainfall is 150 cm or more. During the cool dry season, December to March, dew is a factor in overseasoning of the rust. Susceptible plants tends to tolerate the rust and retain 5 to 15 percent surface infection on older leaves during dry or warm periods. The warm (27 C average annual temperature) and relatively dry (25 to 75 cm annual precipitation) southern slopes of the island which comprise roughly 25 percent of its surface, are free from soybean rust. An increase in the acreage of food legumes, both as garden crops in the highlands and field crops on the northern slopes, will enhance the spread and evolution of this rust on the island.

STUDIES ON THE AERIAL BLIGHT DISEASE OF SOYBEAN AND ITS CONTROL

P. N. Thapliyal, H. S. Verma, R. K. Jain and H. K. Bhadula, Department of Plant Pathology, G. B. Pant University of Agriculture and Technology, Pantnagar, Nainital Distt., U.P., India

Aerial blight, caused by *Rhizoctonia solani* is a serious disease of soybeans during the rainy season in the warm-humid areas like the 'Tarai' region of Uttar Pradesh in India. The disease was originally described from Louisiana. From India it was reported in 1971. Detailed studies were carried out on the disease, its causal organism and on the control of the disease.

In the field, disease appears on at least six-week old plants. All aerial parts of the plant show symptoms as small to large, light or dark brown spots on leaves, petioles, stem and pods. A characteristic cobweb-like mycelial growth that bears sclerotia develops on the plant surface. The infected plants die-back and defoliate. Two morphologically distinct isolates of *R. solani* were found independently associated with the field infected plants. A plant was infected with only one of the two isolates. Both the isolates produced toxic metabolite(s) in culture, which reduced seed germination and seedling root elongation. Metabolite(s) applied to foliage produced necrotic spots. The toxic principle(s) expressed highest toxicity when assayed at 20 C. Autoclaving (15 lb psi for 20 min) reduced but did not eliminate the activity.

Effect of temperature, pH, different media and different carbon and nitrogen sources on growth, sclerotial production and toxic metabolite(s) production by both the isolates were studied. One of the isolates (designated as macrosclerotial) showed a higher rate of mycelial growth and formed round to oval to irregular shaped large sclerotia (1-2 mm) within an incubation period of 84 hr at 30 C. This isolate produces more toxin than the other isolate (designated as microsclerotial). The microsclerotial isolate formed round to oval shaped smaller sclerotia (less than 1 mm) which are formed within 5 to 6 days of incubation at 20 C, however, best mycelial growth was obtained at 25 C. The macrosclerotial isolate was more sensitive to Brassicol and the microsclerotial was more sensitive to Captafol. Both were equally sensitive to benomyl.

Disease could effectively be controlled under field conditions by soil treatment with brassicol (25 ha) plus benomyl seed treatment (0.2 percent) plus two benomyl foliar sprays (0.05 kg/ha).

CALCIUM NUTRITION IN RELATION OF SEVERITY OF ANTHRACNOSE ON SOYBEAN

Onkar D. Dhingra and Luiz A. Maffia, Setor de Fitopatotologia, Universidade Federal de Vicosa, 36570 - Vicosa, MG - Brasil

Soybean seedling blight caused by *Colletotrichum dematium* var. *truncata* causes serious stand losses in soybean growing areas of Minas Gerais. The soils in this area are acidic (pH 5.0 - 5.4) and very poor in calcium (about 0.6 to 1.7 mg eq./100 cc). Lime tests were being conducted but severe seedling blight resulted in total stand loss in plots that were not limed. This study was done in the greenhouse to determine if the calcium content in the soil had some effect on this seedling blight. Sand was washed and amended with Hoagland solution containing calcium (calcium nitrate) levels ranging from 0 to 1000 ppm. The nitrate level was kept constant in all treatment by adding required amounts of potassium nitrate or sodium nitrate. Initially 100 seeds of cultivar 'UFV-1' were planted, which were then thinned to 70 seedlings. Fifteen day old seedlings were inoculated with conidia (10^5/ml) of *C. dematium* var. *truncata*. Uninoculated seedling served as control. Inoculated or uninoculated seedlings were incubated in a mist chamber at 22 - 23 C for 4 days, and then transferred to greenhouse benches. Disease index (DI) ratings were done 10 days after inoculation. The calcium content in plants was determined with an atomic absorption spectrophotometer. The DI and percentage mortality (PM) was significantly higher in plants not supplied with calcium. DI dropped significantly with increased calcium content up to 300 ppm, leveling off at higher concentrations. This correlated with the calcium content of the plants. The calcium content of seedlings that were not supplied with calcium was 0.5 and 1.0 percent when supplied with 300 ppm calcium. In plants treated with 400 - 1000 ppm calcium, the calcium content varied between 1.5 to 1.6 percent. The calcium content of the plant was inversely correlated with severity of anthracnose.

SOYBEAN MOSAIC VIRUS AND ITS EFFECT ON SOYBEAN SEED PRODUCTION

E. K. Allam, Elbagoury H. Olfat and Arif M. Nagwa, Faculty of Agriculture Ain Shams University, Shobra, Cairo, Egypt

Soybean mosaic virus (SMV) had a thermal inactivation point of 63 C, dilution end point of 10^{-3}- and *in vitro* longevity of 72 hours at room temperature. The virus had a limited host range within the family leguminosae.

Soybeans inoculated with the virus had about 50 percent less pods, smaller seeds, significantly fewer smooth seeds, and more small shriveled and mottled seeds than healthy plants. No small shriveled and mottled seeds were obtained from healthy soybean plants.

Germination percentage was 76.1 percent for seeds obtained from plants inoculated at the seedling stage compared with 92.7 percent for healthy seeds.

Virus seed transmission was 52.9 and 32.7 percent for seeds of plants inoculated at the seedling and flowing stages, respectively. Smooth seeds produced from plants inoculated at flowering were free of SMV. Seed transmission occurred in 40.5 percent of the mottled and 20.3 percent of the small shriveled N.C. Hampton seeds.

Oil content was 16.6 and 20.7 percent for mottled seeds of plants inoculated at the seedling and flowering stages, respectively. The oil content of healthy seeds was 22.7 percent. Total nitrogen content of SMV infected seeds was significantly less than in the healthy control plants.

IDENTIFICATION OF SOYBEAN GENOTYPES WITH RESISTANCE OR IMMUNITY TO INFECTION BY OR SEED TRANSMISSION OF SOYBEAN MOSAIC VIRUS

Glenn R. Bowers, Robert M. Goodman and E. H. Paschal, II, Department of Plant Pathology and International Soybean Program (INTSOY), University of Illinois, Urbana, Illinois 61801 and Texas A & M University Agricultural Research and Extension Center, Box 784, Route 5, Beaumont, Texas 77706

Four hundred topically-adapted soybean lines (maturity groups VIII, IX, and X) and 497 lines from temperate maturity groups (II and III) were screened for resistance to the Illinois severe isolate of soybean mosaic virus (SMV-I1-S) and for seed transmission. Field grown soybeans at Isabela, PR (18°N) or Urbana, IL (41°N) were manually inoculated at the primary leaf stage with SMV-I1-S. Seeds harvested from inoculated, infected plants were tested for seed transmission incidence in sand benches in the greenhouse; infected seedlings were detected by observation of symptoms and infectivity indexing. In 1976, all 897 lines were tested with 200 seeds per line. In 1977, 98 tropical lines and 35 temperate lines which had no transmission in 1976 tests were tested with 1000 seeds per line.

Lines that were not infected by virus inoculation in the field in 1976 were tested further in the greenhouse with several SMV isolates to determine if any were resistant or immune to SMV. Of 897 lines tested, tropical lines Buffalo (P.I. 424.131), P.I. 324.068, P.I. 341.242, and P.I. 424.131 proved to be resistant to SMV-IL-S, Ross's blister strain and the ATCC isolate.

Lines with apparent nontransmission of virus through seed and their maturity group numbers were: FC 31.678 (III), P.I. 60.279 (II), P.I. 68.680 (II), P.I. 70.019 (III), P.I. 70.036 (II), P.I. 88.303 (II), P.I. 91.115 (II), P.I. 92.684 (II), P.I. 92.718-2 (III), P.I. 203.406 (VIII), P.I. 240.664 (X), P.I. 325.779 (IX), P.I. 360.835 (II), Arisoy (VIII), Cloud (III), Manchu 2204 (III), Merit (0), Mukden (II), and Virginia (IV).

YIELD LOSSES CAUSED BY BACTERIAL TAN SPOT

J. M. Dunleavy, USDA/SEA, Department of Botany and Plant Pathology, Iowa State University, Ames, Iowa 50011

Bacterial tan spot of soybeans was observed in Iowa from 1975 through 1978, and has been reported from seven counties. The disease is caused by *Erwinia herbicola*, a yellow-pigmented bacterium. Bacterial tan spot differed from bacterial blight and bacterial pustule of soybeans in that leaf lesions continued to enlarge, eventually involving large segments of leaf tissue. Leaflets with large or multiple lesions usually fell from the plants.

Two cultivars, Clark and Clark 63, were tested for possible yield loss caused by *E. herbicola.* Clark 63 is resistant to bacterial pustule disease and therefore might be expected to have resistance to bacterial tan spot. Plants were grown in rows 3 m long, spaced 1 m apart, in randomized blocks with six replications. Each cultivar was grown in three row plots, and the center row was harvested for yield. Bacterial inoculum was prepared by expressing crude sap from leaves infected with *E. herbicola.* Sap was diluted 1/10 with 0.01 M phosphate buffer at pH 7.0. Plants were inoculated when the second trifoliate leaves were unrolling. A bactericidal spray containing 6 ml/liter of 54 percent suspension of copper salts of fatty and rosin acids (4 percent metallic copper) and 0.3 ml/liter of 77 percent phthalic glyceryl alkyd resin as a spreader-sticker was applied to plants in noninoculated control rows until run off, once a week, from time of inoculation until plant maturity.

There was good natural spread of bacterial tan spot in inoculated rows throughout the season, and the disease was present on some leaves at node 20 near the plant apex at plant maturity. The disease did not spready to control rows. Clark and Clark 63 plants in sprayed, control rows yielded 8.9 and 8.0 percent (significant at the 5 percent level) more, respectively, than the corresponding inoculated plants. There was no significant difference in yield (at the 5 percent level) between Clark and Clark 63 in either the inoculated or control rows.

SOYBEAN WILT CAUSED BY *NEOCOSMOSPORA VASINFECTA*

Peter O. Oyekan, Institute of Agricultural Research and Training, University of IFE, Moor Plantation, P.M.B. 5029, Ibadan, Nigeria

Soybean has been grown in Nigeria commercially on a small scale mainly in the Yandev area of Benue State. Very little is known about disease problems of this crop in Nigeria although previous attempts to grow it as a commercial crop in the southwestern part of the country failed.

Importance of diseases in soybean production in Nigeria started to unfold only recently as work to select adaptable tropical varieties for the area progressed. During 1977, soybean grown at Moor Plantation, Ibadan, was attacked by a wilt disease caused by *Neocosmospora vasinfecta.* Infected plants first showed general chlorosis accompanied by drooping of foliage and subsequent withering and death of infected plants. Depending on the stage of development when external symptoms of wilt were manifested, infected plants either produced poorly filled pods or none before drying up. Roots of infected plants appeared normal but the stele along lower parts of the roots always had reddish-brown discoloration which often did not extend above the hypocotyl.

N. vasinfecta was readily isolated from discolored portions of infected roots. When ten-day old soybean seedlings (cv. Bossier) were inoculated by dipping their roots in conidial suspension of the isolate and then replanted in steam pasteurized soil, typical symptoms of wilt as observed in the field developed in the greenhouse.

AN EFFICIENT AND UNBIASED SURVEY METHOD FOR BROWN STEM ROT OF SOYBEANS

H. Tachibana and G. D. Booth, USDA/SEA and Iowa State University, Ames, Iowa

Statewide surveys of brown stem rot (BSR) incidence were made in Iowa in 1966 and 1972. Severity of the disease was greatest in southern Iowa, but the disease was economically important throughout the state. Because recent cursory field surveys indicated that BSR was increasing in northern Iowa, an efficient and unbiased survey method for BSR was designed and tested in northern Iowa counties in 1977. The method emphasized completely random selection of sample collection sites. Each site was preselected by drawing two numbered balls corresponding to specific section coordinates to determine the section from which plant samples were to be obtained. Selection of fields within sections and locations within fields were by additional specific detailed instructions provided the collectors. One sample was collected from each county for each 4,050 ha of soybeans grown in 1976. Specimens were collected from a total of 290 fields by cooperation County Extension Directors of each county. Specimens were subsequently picked-up and examined by plant pathologists at Ames. The counties in the survey were divided into three groups. For each group, specimens were collected in the field the first day, picked-up at county centers and assembled at Ames the second day, and examined for BSR on the third day. Incidence of BSR in individual plants was determined using current standard methods used in selection for BSR resistance. The reported method furnished a more accurate estimate of disease over a wider area than would have been possible otherwise with available research funds and personnel. The results of the cooperative effort further familiarized extension and research personnel with the current status of BSR in the state. BSR was found in 94.5 percent of the fields sampled in northern Iowa. Over the entire area, the disease occurred in an average of 38.0 percent of the plants sampled, with diseased plants showing an average of 35.0 percent of the stem infected. These figures are the highest recorded for the disease in Iowa to date.

THE USE OF PARAQUAT TO DETECT LATENT COLONIZATION BY FUNGI OF SOYBEAN STEM AND POD TISSUES

R. F. Cerkauskas, J. B. Sinclair and S. R. Foor, Department of Plant Pathology, University of Illinois, Urbana, Illinois 61801

Paraquat (1,1'-dimethyl-4,4'-bipyridinium dichloride) is a herbicide used as a harvest aid for many crops. *Phomopsis* and *Alternaria* spp. were recovered more often from seeds of Paraquat-sprayed soybean plants than from nonsprayed plants in 1977. We wished to determine if Paraquat could be used as an aid in detecting latent fungal colonization in soybean stems and pods.

In 1978, 'Bonus' soybeans were harvested weekly beginning July 10 (40 days after planting, stage V8) through August 20 (81 days after planting, stage R5.5). Stems from randomly selected plants were cut into 4-cm lengths, washed in running tap water for 5 to 6 hours, dipped in 95 percent ethanol, rinsed in 0.5 percent NaOCl (10% Clorox) for 4 minutes and finally rinsed in sterile distilled water for 1 minute. Twelve randomly selected pieces from these stems were dipped in a filter-sterilized water solution of Paraquat (1:40 dilution of formulated product) for 45-60 seconds. Twelve comparable stem pieces were nontreated (control). Stem pieces were incubated 4 days in sterile culture plates on moist filter paper in continuous light at 25 C. Nontreated stem pieces were incubated an additional 6 days. Mycelial growth (primarily *Fusarium* and *Phomopsis* spp.) was rated on a scale of 1 to 5. A significantly ($P = 0.05$) greater percentage of stem pieces had abundant mycelial growth when treated with Paraquat than nontreated at each harvest date after 4 days incubation and in four of five sampling dates when compared to controls incubated an additional 6 days.

Pods and stems of 'Amson' and 'Hitatsa' inoculated with *Cercospora kikuchii* (5,000 conidia/ml) were harvested from field plants and treated as previously described but with a 15 min. tap water wash and no ethanol dip. Significantly ($P = 0.05$) more *Cercospora* lesions were recorded on stems and pods from Paraquat-treated than from nontreated tissues.

Paraquat did not improve recovery of *Alternaria* spp. from pods and stem tissues of 'Hawkeye.'

Treatment of soybean pods and stems with Paraquat significantly increases the rate and ease of recovery of various fungi and may aid in detecting latent colonization of soybean stem and pod tissues.

EFFECT OF POD INOCULATION WITH *PHOMOPSIS* SP. ON SEED GERMINATION OF TWO SOYBEAN CULTIVARS

M. A. Ellis, Department of Crop Protection, University of Puerto Rico, Mayaguez, Puerto Rico 00708; O. Zambrano, Instituto Nacional de Investigaciones Agropecuarias, Estacion Experimental Portoviejo, Portoviejo-Manabi, Ecuador; and E. H. Paschal, Texas A & M University Agricultural Research and Extension Center, Route 7, Box 999, Beaumont, Texas 77706

One of the factors limiting the increased production of soybean [*Glycine max* (L.) Merr.] in the tropics is the production of high quality seed for planting. Research conducted by the International Soybean Program (INTSOY) at the University of Puerto Rico has demonstrated that a great deal of variability does exist between soybean cultivars for seed quality characteristics under tropical conditions. Under identical conditions, certain cultivars have the ability to resist decreases in seed viability and increases in seed infection by fungi under delayed harvest conditions, while others lose viability rapidly and may become 100 percent infected by fungi. Seedborne fungi such as *Phomopsis* sp. have been associated with reduced soybean seed germination in-vitro and reduced emergence in the field. The objective of this research was to observe the reaction of two selected soybean cultivars to seed infection by *Phomopsis* sp.

The cultivars PI 205912 and Hardee were selected for use in this study. PI 205912 has shown good seed quality characteristics (high viability and low incidence of fungi)under delayed harvest conditions. Hardee has shown poor seed quality characteristics (low viability and high incidence of fungi) under identical conditions. Both cultivars reached maturity at the same time (109 days). Attached pods of both cultivars were inoculated in the greenhouse at four different stages of development with isolates of *Phomopsis* sp. recovered from infected soybean seeds. Pods were inoculated by immersion in a suspension of mycelium and spores of *Phomopsis* sp. Uninoculated pods served as controls. All pods were harvested at one week after maturity and their seeds assayed for incidence of *Phomopsis* and germination in-vitro.

Phomopsis was not recovered from seeds of uninoculated pods of either cultivar. The mean percentage seed infection and germination in-vitro of seeds were 61 amd 71 percent, respectively, from innoculated pods of PI 205912, and 67 and 11 percent, respectively, for Hardee. The percentage seed germination in-vitro of seeds from uninoculated pods of PI 205912 and Hardee was 88 and 79 percent, respectively. There was no significant difference in the incidence of *Phomopsis* in seeds from inoculated pods of both cultivars; however, seeds of PI 205912 which were infected by *Phomopsis* had a significantly higher in-vitro germination percentage than infected seeds of Hardee. This data suggests that PI 205912 may possess some form of tolerance against seed or seedling death incited by *Phomopsis* sp.

SOYBEAN SEED DETERIORATION IN THE TROPICS: I. THE ROLE OF PHYSICAL FACTORS AND PATHOGENS

H. C. Wien, B. Ndimande, P. R. Goldsworthy, International Institute of Tropical Agriculture, PMB 5320, Ibadan, Nigeria

Three experiments were conducted to determine whether the rapid loss in soybean seed viability in the lowland humid tropics is due to physical factors and/or to pathogens. In the first two, benomyl fungicide was applied as a foliar spray weekly from flowering to four cultivars of contrasting storability. The seeds were harvested either promptly or with a delay of two weeks from maturity. Plants in Experiment 1 matured under dry conditions and low fungal incidence so that harvest delay and benomyl treatment had no significant effect on seed pathogen incidence or germination at harvest. In Experiment 2, seeds matured in moist conditions, and delayed harvest reduced viability by 25 percent, with significant cultivar X harvest time interactions. Seeds from plants treated with benomyl had 2 and 9 percent higher viability than from untreated plants for prompt and delayed harvest, respectively. Incidence of seed-borne *Aspergillus, Cercospora kikuchii, Macrophomina phaseoli, Phomopsis,* and *Colletotrichum truncatum* at harvest was slightly reduced by benomyl application.

To determine the effect of benomyl treatment on rate of decline in seed viability in storage, seeds from Experiment 1 and 2 were stored at 80 percent RH and 28 or 35 C and sampled for seed pathogen incidence and germination. Seeds from benomyl-treated and untreated plants declined in viability at equal rates. In spite of low incidence of seed-borne microorganisms in Experiment 1, viability declined rapidly at 35 C.

Dusting of benomyl on seed of Bossier soybean equilibrated to 10, 13 or 16 percent moisture content before the start of storage at 28 C completely inhibited surface fungal growth but did not influence rate of viability decline. Rate of viability decline varied directly with seed moisture content, supporting the concept that physical factors alone can reduce viability of seed in storage. The role of pathogens in seed deterioration may frequently be secondary.

SOYBEAN SEED DETERIORATION IN THE TROPICS: II. VARIETAL DIFFERENCES AND TECHNIQUES FOR SCREENING

H. C. Wien, P. R. Goldsworthy, E. Kueneman, International Institute of Tropical Agriculture, PMB 5320, Ibadan, Nigeria

Fifty lines of Southeast Asian or of US origin were field-planted in four successive seasons after periods of storage ranging from two to nine months under ambient (27 + 2 C, 80 percent RH) conditions to identify soybean lines that can maintain good germinability after prolonged storage under the conditions of high temperature and relative humidity prevalent in the tropics. We hoped to determine differences in preharvest deterioration by harvesting the lines at maturity or with a delay of two weeks. Results indicated that storage duration was the major factor influencing germinability, while delayed harvest had less, but a more variable effect. Consistent differences in germinability were found among lines with some small seeded lines from Southeast Asia maintaining more than 50 percent germination after 8 months storage under ambient conditions. Accelerated aging (40 C, 100 percent RH for 3 days, followed by germination test) did not correlate well (r = + 0.05) with storability of lines under ambient conditions, because of profuse fungal growth and subsequent cross contamination.

Storage tests with eight lines of contrasting storability identified in the preceding experiments were produced in one-third the time by lowering the RH of the standard aging test from 100 to 60-80 percent. This greatly reduces fungal growth and slows the aging process. The close monitoring of germination of standard lines included in each test insures the attainment of reproducible stress levels. Seed of F_3 and subsequent generations from crosses of high yielding parents with those of outstanding storability are being screened with the modified aging procedure to combine seed longevity and high yield.

STUDY ON THE DISEASE TRANSMISSIBILITY BY SOYBEAN SEEDS [*Glycine max* (L.) MERRILL]

Walner da S. Fulco, Maria Regina Camargo, Agronomic Research Institute (IPAGRO), Goncalves Dias 570, P. Alegre 90.000, RS, Brazil

Seed health testing was conducted at the Phytopathology and Seed Technology Laboratories (IPAGRO, 1977), to evaluate the bacterial and viral incidence and to estimate the transmissibility of these pathogens in soybean. The following soybean cultivars from each of six counties of the Rio Grande do Sul State were assayed: Bossier, Bragg, Davis, Hardee, Hood, Ias 4, Parana, Perole, Planalto, Prata, Santa Rosa and Sulina.

The blotter-method was utilized in the testing of seeds for fungi, direct planting to discover the presence of viruses and bacteria, and sowing in plastic dishes for bacterial determination. The sample was composed of 50 seeds per cultivar and per local, with 3 replications.

In the event of fungi detection, the seeds were observed at the 8 days after incubation at 20 C. In the test for virus transmission, the seeds were separated into two lots: seed coat without blemishes and seed coat with blemishes. The virus detection was conducted by direct planting. Plants with symptoms were submitted to serological methods and plants free of symptoms were assayed through inoculation in susceptible plants.

The different techniques indicated that the more prevalent pathogens on the assayed cultivars were: *Rhizoctonia* sp., *Cercospora kikuchii, Fusarium* sp., *Phomopsis* sp., *Colletotrichum* sp., *Cercospora sojina, Dreschslera* sp., *Diaporthe phaseolorum* var. *sojae, Helminthosporium* sp., and *Xanthomonas* sp.

Rizoctonia sp. and *Cercospora kikuchii* were isolated from all cultivars and locals. Fewer *Fusarium* sp. and *Phomopsis* sp. were isolated but were still present at all levels.

Attention must be paid to the harvest, handling and storage which affect the seed quality, particularly in tropical climates.

FACTORS AFFECTING INOCULUM POTENTIAL OF *MACROPHOMINA PHASEOLINA*

T. D. Wyllie and S. Gangopadhyay, University of Missouri, Columbia, Missouri

Macrophomina phaseolina causes charcoal rot of soybean and is worldwide in distribution. The amount of damage caused to soybean plants is, in part, a function of the numbers of sclerotia per unit volume of soil and the endogenous and exogenous carbohydrate that is available for germination, growth, and infection. In order to determine the amount of carbohydrate necessary for these activities it was essential to remove the available CHO before resupply could be quantitated. Failure to germinate was used as the measure of depletion of CHO. Exogenous CHO's were removed by washing the propagules in sterile demineralized water until the detectable carbohydrate yield in the wash water as 10^{-10} M. Endogenous CHO was removed by ultrasonication in 0.1% NaOCl and successive aqueous centrifugation. Leaching was terminated when the yield of CHO equaled 10^{-10} M. Leachable CHO reserves, however, were replenished upon storage of the sclerotia for up to 21 days, thus enabling a quiescent propagule to overcome fungistasis apparently through the conversion of lipid reserves to utilizable CHO. This information offers an explanation for the lack of uniformity of germination of sclerotial propagules within a population.

PRODUCTION OF HIGH QUALITY SOYBEAN SEED IN FLORIDA AS AFFECTED BY RAINFALL DISTRIBUTION, MATURITY AND HARVEST DATE, AND STORAGE CONDITIONS

L. J. Alexander, P. Decker and K. Hinson, Department of Plant Pathology, University of Florida, Gainesville, Florida 32611 and ARS, USDA, Department of Agronomy, University of Florida, Gainesville, Florida 32611

In 1970, it was deemed important to determine whether good soybean seed of determinate types could be produced in Florida. This question was of primary importance for several reasons: seed from neighboring states was frequently poor, the danger of importing cyst nematode, and locally produced seed frequently resulted in poor crops. Fungi are the principle causes of poor seed quality and germinability. The maturity-to-harvest period was studied to determine effects of time interval between harvest and maturity, rainfall, and storage conditions affecting fungus infection of seeds and reduced germination.

Determinations of seed infections were made by culturing 20 seeds on lima bean agar after surface disinfection. Germination determinations were made by placing 50 seed on moist filter paper in 150 mm petri dishes. Incubation periods were 5 to 7 days. Moisture determinations were made in a vacuum oven. In 1972 fields were relatively wet after soybean maturation. Five harvests were made between October 30 and November 29. At first harvest a relatively low percentage of beans was infected and germination was excellent. At each succeeding harvest, the infection percentage increased and germination decreased. At last harvest, most beans were infected and few germinated. The exception was the very late maturing Jupiter. Rainfall in 1974 was much less after beans matured. Forrest matured immediately after a prolonged wet period, and seed infection was high and germination poor. Later maturing varieties became infected slowly with fungi and germination remained good. Apparently overcast skies with some rain accelerates seed infection more than heavy rains followed by clearing skies. The results of 1973 and 1975 were similar to 1972 and 1974 and are omitted.

Using purple stain of seed, as a model, it was determined that gross seed infection was readily visible, but it was impossible to determine accurately the amount of seed infection, except by cultural methods. The average percent visual infection was approximately 2 percent whereas that determined by culture usually was several times higher. The greatest single difference was with Centennial, Oct. 16 harvest, where 1 percent infection was observed but 15 percent found by cultural methods. Rainfall by months over a 40-year period was plotted for four areas of Florida where soybeans are most frequently grown. The rainfall pattern was similar for the four areas. Of importance was the fact that usually a dry period occurred during late October and early November, the important time for harvest. These results indicated that high quality seed can be produced in Florida.

IDENTIFICATION AND DETERMINATION OF THE CRITICAL PERIOD FOR TIME OF INOCULATION OF TOBACCO RING SPOT VIRUS IN SOYBEAN VARIETY JUPITER IN THE SOUTHERN AREA OF TAMAULIPAS, MEXICO

M. de los A. Pena del Rio, Agric. Exp. Sta. "Los Huastecas", CIAGON-INIA-SIRH, Tampico, Tamaulipas, Mexico

In the southern area of the State of Tamaulipas, Mexico, soybean acreage has increased from 300 ha in 1968 to 100,000 ha in 1977 with the potential for more than 400,000 ha of soybeans depending on general rainfall conditions. In 1977 and 1978 more than 10 percent of the soybean acreage was suspected of having 100 percent infection from tobacco ring spot (bud blight) virus (TRSV). In the fall of 1977, collections were made in fields with 100 percent "evergreen" soybean plants. The disease was transmitted to healthy and young soybean plants under field conditions, with transmission varying from 60 to 80 percent. From the inoculated plants, the disease was transmitted to TRSV and other virus diagnostic hosts. The symptoms produced in the diagnostic hosts were typical for TRSV. Readings were obtained from 3 to 20 days after inoculation, depending on the diagnostic host. Some physical properties of the TRSV were determined such as thermoinactivation and longevity in vitro with positive results. In 1978 serological tests reconfirmed the presence of TRSV. During the 1978 growing season, under field conditions, an experiment was designed to determine the critical infection period in soybean variety Jupiter. There were four periods of inoculation from tobacco to soybean, one during the vegetative period (V6), and three during the reproductive period: during flowering (R2), during pod formation (R3) and during pod fill (R5). The transmission efficiency was 95 percent at V6 and 19 percent in R2. Yield reductions at V6 and R2 were 95 percent and 27 percent, respectively. There were no significant ($P > .05$) yield reductions for infection at R3 and R5. These studies suggest that soybeans should be protected from virus infection in dry years during the vegetative, bloom and pod-fill stages (V1-R3).

MODELING SOYBEAN SYSTEMS

Invited Papers

SIMULATION FOR RESEARCH AND CROP MANAGEMENT

D. N. Baker, USDA, SEA, Crop Simulation Research Unit, Mississippi State, Mississippi 39762

This paper discusses what are classified as predictive, discrete, process oriented, dynamic simulation crop models. A brief history of this form of modeling in crop production research is offered with objectives and methodology. The subroutine structure is identified and justified. Technical problems including validation and the data base for rate equations and the quality and availability of input data are reviewed. This is related to activities in USDA, NOAA, NASA and the state experiment stations.

An assessment of progress to data is made, together with an identification of the size of the effort and its rate of growth, its strengths and its weaknesses. Applications discussed include large area yield forecasting, crop management decision making, crop system analysis and design. A discussion of the current computer requirements including core size, execution time and language, and a glimpse at the future are provided.

SIMULATION OF INSECT DAMAGE TO SOYBEANS

W. G. Rudd, Department of Computer Science, Louisiana State University, Baton Rouge, Louisiana 70803

A simple canopy model for soybean above-ground dry matter production has been developed for use in insect pest management modeling research. The model is based on an empirical relationship between leaf area and dry matter accumulation rates. The model outputs include dry weights of stems, leaves and fruit as functions of days after planting. Validation runs show good agreement between model outputs and experimental data for several soybean varieties under different growing conditions.

Experiments with simulated mechanical defoliation indicate that some "compensation" occurs if the plant is heavily defoliated after podset. The model agrees well with experimental results when compensation is included. The model also responds correctly to simulated podfeeding and to variations in planting dates. The model is now in use in pest management modeling studies.

SIMULATION OF VEGETATIVE AND REPRODUCTIVE GROWTH IN SOYBEANS

R. B. Curry, Department of Agricultural Engineering, Ohio Agricultural Research and Development Center, Wooster, Ohio

The current state of the art of the application of simulation to the understanding of soybean vegetative and reproductive growth will be discussed. Major emphasis will be placed on SOYMOD/OARDC, a dynamic soybean growth and development simulation. Research on soybean simulation at other locations will be summarized. SOYMOD/OARDC is a modular structured computer program written in FORTRAN IV consisting of a mainline program and 22 subroutines. Subroutines include mathematical representation of processes and also provide for interaction between the processes and the feedback to the processes. Processes include photosynthesis, respiration, nitrogen assimilation, translocation, storage, carbohydrate and nitrogen partitioning, growth, flowering, podfill, and senescence, along with a soil plant water balance system. The process representation in the subroutines is mechanistic, based on the current state of knowledge obtained from local experiments or from the literature. Inputs include weather data, plant and soil parameters. Output data are available in several formats to meet the needs of the user. The simulator has been verified and validated. Examples of potential use in research will be presented.

APPLICATION OF MODELING TO IRRIGATION MANAGEMENT OF SOYBEAN

J. W. Jones and A. G. Smajstrla, Agricultural Engineering Department, University of Florida, Gainesville, Florida 32611

The use of models to relate crop water use and irrigation management is discussed in this paper. Early modeling efforts for irrigation scheduling were mass balance approaches using soil water storage and depletions from storage by evapotranspiration (ET). ET models were developed to predict water use as a function of climatological parameters and empirical crop coefficients. Coefficients were assumed to be constant for monthly growth periods, and soil water was assumed to be equally available to the plant between field capacity and the permanent wilting point. Crop growth was assumed to be unaffected by soil moisture in the available range.

A second level of sophistication in modeling recognized the effects of soil moisture status on crop growth and yield. Models are described which relate water use and final crop yield to soil moisture status at various stages of growth. Water is no longer considered equally available over a wide range of soil moisture conditions. Examples of the use of such models for determining irrigation management strategies and for scheduling of irrigations are discussed.

A third level of sophistication is discussed. This approach consists of the development and use of a dynamic crop growth model for simulating growth and ultimately yield as a function of dynamic soil, plant and atmospheric conditions.

Requirements for development of each type of model and advantages and disadvantages of each approach are discussed. A framework is outlined for application of a crop response model, coupled with soil water and evapotranspiration models, and economic factors for scheduling irrigation in soybeans.

THRESHING AND SEPARATING PROCESS—A MATHEMATICAL MODEL FOR SOYBEANS

V. M. Huynh and T. E. Powell, Engineering Department, White Farm Equipment, Brantford, Ontario

A mathematical model is developed to describe the process of threshing and separating soybeans in a conventional cylinder and concave thresher. Stochastic concepts are used to quantify the behavior of crop material during threshing. Three important physical phenomena are identified: the detachment of seeds from stalks, the penetration of free seeds through the straw mat and the subsequent passage of seeds through the concave grate. Each of these phenomena is presented by a probabilistic distribution function whose parameters are related to operating variables, thresher design parameters, and crop conditions. These are then related to threshing loss and separation efficiency. Threshing performance of the cylinder and concave thresher is also evaluated in terms of threshing horsepower demand and seed damage. Theoretical results indicate the trends for the effect of soybean threshing parameters on threshing performance.

SIMULATION APPLICATIONS IN SOYBEAN DRYING

J. H. Young, North Carolina State University, Raleigh, North Carolina 27650 and R. N. Misra, Ohio Agricultural Research and Development Center, Wooster, Ohio

Finite element procedures are used to simulate the response of soybeans to various drying conditions. Procedures have been developed which allow the simultaneous prediction of drying rates and shrinkage of beans. Observations of cracking percentages for soybeans dried under different conditions have been compared with stress predictions based on elastic material properties. Results indicate that soybeans are viscoelastic materials and further work is underway to predict stresses using viscoelastic properties and thus to provide techniques for optimizing the drying conditions used.

MODELING SOYBEAN INSECT POPULATION DYNAMICS

W. G. Rudd, Louisiana State University, Baton Rouge, Louisiana

Abstract not available at press time.

REGIONAL

Contributed Papers

POOR SEED QUALITY—A MAJOR LIMITING FACTOR IN SOYBEAN PRODUCTION IN INDIA

J. N. Singh and G. D. Gross, G. B. Pant University of Agriculture and Technology, Pantnagar, India

Several exotic varieties of soybean grown in India give very high yield but poor quality of seed. Production of poor quality seeds, rapid loss of viability during storage, and poor and erratic field stand of the crop are the major problems in soybean cultivation in India. A series of experiments therefore were conducted in the field, laboratory and phytotron to find out the reasons for poor quality and storability of soybean seeds. Indigenous varieties (T1, T49 and Kalitur) produced better seeds of better quality, storability and vigor. Physical analysis revealed a larger percentage of seeds with damaged testa (wrinkled, ruptured, or shrivelled) and such seeds lost viability fast and had low seed vigor. Sound seeds with 8 percent moisture maintained more than 80 percent germination by the next planting season. Prevailing high temperature and high relative humidity during crop growth produced seeds with damaged testa. Crops planted late in September/October or on high elevation (1000-1500 m above sea level) produced better quality seed and maintained higher viability and vigor in storage than the seeds obtained from June/July (normal time) plantings in the plains. Mature pods harvested and stored at 30 C temperature and 75-85 percent relative humidity showed considerable reduction in seed viability and vigor only after 15-30 days of storage. The pathological examination of harvested seed revealed the presence of large numbers of pathogens on the seed surface. Prevailing storage temperature, especially from May to June, and high relative humidity were primarily responsible for deterioration of seed quality. Among the various packaging materials tested for storing the seeds, hessian bags and cloth bags were found to be better than plastic or polycoated hessian bags. Improper mechanical threshing was found to be another important factor affecting the seed quality. Seed treatment with fungicides (Captan and Thiram) invariably enhanced the germination in the laboratory as well as in the field.

DEVELOPMENT OF A SOYBEAN INDUSTRY IN SRI LANKA

E. Herath, W. Wijeratne, C. N. Hittle and J. M. Spata, INTSOY Soybean Program, C.A.R.I., Gannoruwa, Sri Lanka

Experiences of the Sri Lankan Soybean Development Program illustrate the opportunities and problems associated with a broad-based national program of crop production, marketing, processing, and utilization. The International Soybean Program (INTSOY), based at the University of Illinois and the University of Puerto Rico, has assisted with the research and educational component of this program in collaboration with the government of Sri Lanka and several of its agricultural organizations. Support has been provided by the United Nations Development Program (UNDP) and the Food and Agriculture Organization of the United Nations (FAO).

In 1977, there were about 2400 ha of soybeans planted. Farm yields average slightly less than 1100 kg/ha while experimental trials yield as much as 4000 kg/ha. Current production research includes work in breeding, agronomic practices, soil and water management, development of Rhizobium inoculum and plant protection. This research plus extension should increase farm yields appreciably.

The processing and utilization program is an ancillary to the main project and funded by CARE/UNICEF with local counterpart funds. The foreign funds ($227,000 each from CARE and UNICEF) are spent on machinery, equipment and foreign expertise while local funds provide for land, buildings, staffing and infrastructural facilities. The Research and Development efforts of this complex will cater to three levels of technology: 1) *Home Level:* The objective of this activity is to develop processes and products so that soybeans can be processed in individual homes. Processes will be recommended in such a way that they can be carried out with basic facilities available in the average kitchen in Sri Lanka. Ladies from the Extension Division of the Department of Agriculture and non-government organizations are being trained to teach soybean cooking at the home level. 2) *Village Level:* The development of small scale labor intensive industries on soy processing will generate employment and help improve nutrition of the people. Products such as soy meal, snack foods, breakfast foods, sweetmeats and soy beverages are considered appropriate for this level of activity. 3) *Commercial Level:* A food processing specialist has set up a pilot plant. This facility now develops products and processes that will ultimately be taken up by commercial ventures. The plant will develop technology, demonstrate technical and economic feasibility and produce the products in adequate quantity for consumer surveys.

STATUS, PRODUCTION POTENTIAL AND AGRO-GENECOLOGICAL CONSIDERATIONS FOR SOYBEAN IMPROVEMENT IN INDONESIA

S. Somaatmadja and O. O. Hidajat, CRIA, Sukamandi, Indonesia

Being of short duration and having rather wide adaptability, soybean is grown primarily as a secondary crop after rice in double and multiple cropping systems or as an intercrop in Indonesia. Depending upon the availability of irrigation and local rainfall patterns, its cultivation is exerted in different systems which are generally grouped into: 1) simple, 2) semi-intensified, and 3) intensified method of cultivation.

Low yields obtained from local cultivars indicated the need for intensive production research as well as the utilization of better production practices. CRIA has released a number of cultivars with better yield potential and grain quality. These cultivars are products of selection, either from naturally occuring variability or created by limited hybridization and then tested under sole cropping. Although possessing desirable characteristics, these cultivars have not been fully utilized due to their later maturity and unfitness in existing cropping systems.

Susceptibility of present day cultivars to insect pests and lately also diseases, constitutes one of the main constraints to high yield. Breeding for resistance to major insect pests and diseases is high on the priority list of the breeding program, next to other effective control measures. Selection and breeding work are further attempting to incorporate high yield potential, good seedling vigor, early vigorous growth and tolerance to drought stress into elite domestic cultivars. Exotic germplasm is also being used in this breeding program. In addition, selection for genotypes having greater production efficiency under mixed and intercropped systems is receiving proper attention.

CROPPING SYSTEM AND UTILIZATION OF SOYBEAN IN NEPAL

M. P. Bharati, Legume Improvement Program, Agronomy Division, Kathmandu, Nepal

Nepal is divided into four physiographic regions which differ sharply in elevation. This results in much variation in temperature with a short distance from south to north. Soybean production and utilization is ancient in Nepal. Evidence indicates that soybean could have been introduced in Nepal from South China. Soybean is widely grown in the hills of Nepal. Large variation exists among indigenous germplasm. Mixed and intercropping of soybean and maize, and cropping in bunds of rice terraces are common methods of cultivation. Diversity exists in traditional ways of uses of soybean. These range from daily consumption to sacred use in special occasions. Uses include green bean, dried bean and fermented bean.

PROTECTIVE RESEARCH ON SOYBEAN IN THE PHILIPPINES

F. C. Quebral, R. S. Rejesus, D. A. Benigno, O. S. Opina and R. P. Robles, University of the Phillippines at Los Banos, College of Agriculture, Laguna, Philippines

Soybean is threatened by several injurious pests and diseases. About 208 species of insects, 7 species of weeds and 9 different kinds of diseases have been found associated with soybean. Without adequate control measures these pests and diseases could destroy the entire crop or considerably reduce the yield. The concern of this program is to evolve control measures against pests and diseases.

Field testing of pesticides has resulted in the recommendation of 8 insecticides against critical pests such as bean fly, bean aphids, leaf folders, pod borers and green stink bug, 4 herbicides for the control of grass and broadleaf weeds, and 5 fungicides for the control of rust, damping-off and seed borne diseases. Germplasm screening has resulted in the identification of several varieties possessing some degree of resistance to rust, bacterial pustules, mosaic, root and stem rot, and nematodes. UPLB-SY 2 and L-114, two locally developed varieties which are now commercially planted are resistant to bacterial pustules, mosaic, root and stem rot, and root knot nematodes.

SOYBEAN BREEDING IN MALAYSIA

T. C. Yap, Department of Agronomy and Horticulture, University of Agriculture, Malaysia

Due to the lack of locally adapted high yielding soybean varieties with acceptable grain quality there is no commercial production of this crop in Malaysia. A large quantity of the soybeans imported every year are from soybean production countries. Nevertheless in recent years some breeding work and evaluation of imported genotypes have been conducted to develop some good varieties for local production in order to save some foreign exchange. This paper deals with the research activities on soybean breeding carried out in this country.

SOYBEAN SITUATION IN INDIA (RESEARCH, EXTENSION, PROCESSING, UTILIZATION AND POPULARIZATION)

R. N. Trikha, Soya Production and Research Association, Bareilly, U.P. India

Soybean is a newly introduced crop to Indian agriculture. The importance of soybean as pulse, oilseed and food crop in a predominantly vegetarian society like India is obvious. Soybeans have been grown for centuries in hills and some scattered pockets in Central India. However, the varieties grown have generally been of very low yields. Prior to 1965 several attempts to introduce soybean in the plains failed although the soils and climate of the plains are suitable for soybean production. The potential and problems of soybean production in the plains have been described.

Encouraged by the success of soybean production at Pantnagar and Jabalpur, the Indian Council of Agricultural Research, New Delhi, started an Inter Institutional coordinated Soybean Research Project in 1967. A maximum total of 4198 germplasm lines and 5 wild soybean species are being maintained at Pantnagar. Emphasis has been on the development of varieties suitable for different regions of the Indian subcontinent which extends from 8 to 38° N latitude. Experiments have been conducted in different parts of the country to develop production methods for different agricultural conditions. High yields of 25 to 35 quintal per hectare can be obtained with new varieties and recommended cultural practices. Of a large number of farmers in the plains of Uttar Pradesh many were growing soybean with suboptimal profits because of field problems such as poor germination, lack of nodulation in newer areas, mung bean yellow mosaic virus, marketing and utilization. Unfortunately, no new scheme existed to coordinate farmers in marketing, processing and utilization.

EVALUATION OF THE EFFECT OF STORAGE CONDITIONS ON THE QUALITY OF SOYBEAN SEEDS IN EGYPT

O. H. El-Bagoury, M. T. Fayed, and M. T. Hegab, Ain Shams University, Cairo, Egypt

Adequate provisions and facilities for storage of seed are important components of seed production marketing operations in all climatic regions. However, they are essential in sub-tropical and tropical regions because of the general adversity of the climate for storage of seed.

This investigation included two storage experiments in two successive years (1976 and 1977). Each experiment included eight treatments which were the combination of soybean varieties Clark and Boisser, immature and mature dates of harvest and ambient and 10 C storage temperatures. Experiments were conducted in complete randomized design (nested experiment) with three replications. Seeds were stored in the two experiments for one year and the data were obtained every two months. Germination percentage after 8 days from planting, germination rate index, fresh and dry weight of seedlings, hundred seed weight, oil percentage and quantity, oil acid value, iodine value, free fatty acid, value as oleic acid, total soluble carbohydrate percentage and quantity, protein percentage and quantity and seed moisture percentage were the data obtained.

The following data were recorded. The germination percentage, germination rate index, seedling fresh and dry weight increased with decreasing the temperature of storage while seed moisture percentage, total soluble carbohydrate rate, acidity value oil, free fatty acid value as oleic, weight of 100 seeds, iodine number and total protein increased with increasing storage temperature. The germination percentage and germination rate index increased during the first four months of storage and then decreased with increasing storage period. Seed moisture decreased while soluble carbohydrate and acidity oil value increased with increased storage period. Protein and oil percentage were almost stable during storage. Clark and Boisser varieties varied slightly in germination percentage, germination rate index, and 100 seed weight. Storage at 10 C decreased the total soluble carbohydrate, acidity value and number of free fatty acid as oleic in both varieties. The iodine number was increased slightly. Immature seeds did not germinate. Mature seed germination percentage, germination rate index, and fresh and dry weight of seedlings increased at the 10 C storage temperature. Total soluble carbohydrate of both varieties increased with increasing storage periods. Acidity value and number of free fatty acid as oleic increased with storage period. The interaction between maturity and storage temperature did not obviously affect the oil percentage, although oil percentage decreased gradually with increasing storage period. Storing seeds at room temperature increased the weight of 100 seeds, total soluble carbohydrate, acidity value and number of free fatty acid as oleic.

UTILIZATION OF SOYBEAN IN A DEVELOPING COUNTRY LIKE INDIA–AN OPINION

N. C. Mahajan, College of Agriculture Palampur-176062, Himachal Pradesh, India

Soybeans which are an excellent source of quality proteins and fats can go a long way to help feed poor people. However, this is not happening. The excellent work done by the breeders and agronomists is going to waste. The soybean producer sometimes does not find a buyer of beans at a reasonable price. All this is the result of excessive fine quality work done on soybean chemistry and technology of processing. Its antinutritional factors, beany smell, and chalky bitter taste are too well advertised. In India, where soybean is a new crop to be grown on a large scale, people already have prejudices against it. Taking soybean to the processing unit increases the cost of the processed convenient food to a height not within the reach of a poor pocket. The processed soy-based food is only helping the already well-fed people of the society. In India where pulses and legumes are the major source of protein, it should be tried to persuade people to replace pulses with soybean in their meals at least once a week. This will go a long way to help improve dietary standards of the ill fed and also release some pressure from the pulses which are in short supply and their per capita availability becoming more and more limited. In some households the practice of consuming home processed soybean is already underway.

CHARACTERISTICS AND VARIABILITY OF SOYBEAN YIELD COMPONENTS IN DIFFERENT ECOLOGICAL CONDITIONS: IMPORTANCE OF FRUIT ABORTION

J. Puech, M. Hernandez, J. Salvy, Station d'Agronomie, Institut National de la Recherche Agronomique, Castanet Tolosan, France

With the object of evaluating some limiting factors of soybean yields in France (for indeterminate type, especially group 0, I and II) the influence of various experimental (trophic and edaphic) conditions were studied. The measurements concerned the total number of fruit-bearing organs set or not set during the vegetative cycle and the quantity of seed produced at the different nodes of the main stem. The principal results are as follows. 1) The number of pods presents a relation of the hyperbolic type with the number of plants per area unit (trials between 5 and 80 plants/m^2). 2) The number of fruit-bearing organs shows a maximum point at the beginning of the filling phase of seeds into the first formed pods. The value of this maximum is greater (with a same density of plant/m^2) when the photosynthetic conditions and nitrogen supply (mineral or symbiotic) are increased. The minimum is obtained with strong water stress. 3) In the different treatments (light deficient or excess, growth regulators, abscission of organs, water stress, etc.) the abortion rate is important (60 to 80 percent), but it can be changed. It depends upon the physiological stages of organs and on their position on the main stem, branches or nodes. There is a period of sensitivity for the young pods when they are in competition with 30 to 40 percent of ripening pods. In addition, there is a seed abortion in the pods which concerns especially the seeds nearer the peduncle. These seeds have generally a length between 2 and 3 mm. The fill rate of pods varies in these trials between 70 and 95 percent of available cells.

Soybean is characterized by several gradients: number of pods, weight of the seed in the pods (generally the seed nearer the peduncle is the lightest) and along the main stem for instance. The middle part of the plant has the best productivity but the amplitude between the top and the base of the stem can decrease with an increase of photosynthesis (light supply) and nitrogen supply at specific phases. Lastly, the abortion rate of fruit bearing organs is important. However, the abortion rate may be influenced by plant metabolism, translocations, and concurrence between organs.

SOYBEAN [*GLYCINE MAX* (L.) MERR.] RESEARCH IN THE SUBHUMID TROPICS OF WEST BENGAL, INDIA

B. N. Chatterjee and M. A. Roquib, Faculty of Agriculture, Bidhan Chandra Krishi Viswa Vidyalaya, India

Soybean cultivation prior to 1965 remained confined mainly to the high hills during monsoon. In collaboration with the All-India Coordinated Research Project on Soybean sponsored by the Indian Council of Agricultural Research, extensive work has been done on various aspects of the crop in this area since 1967. The work has been summarized in brief. 1) More than 700 germplasm collected from all over the world through Plant Introduction Division, Indian Agricultural Research Institute, were screened. A number of varieties (Bragg, Improved Pelican, Lee, and Soyamax) have been identified to do well in monsoon season giving about 2.5 to 3.0 t/ha of grain. 2) Monthly sowing of some of these varieties revealed that Bragg and Improved Pelican can be sown in late winter (January) and can give comparable yield of grain with slight modification of agronomical practices. 3) The management practices of the two high yielding varieties (Bragg and Improved Pelican) have been studied for monsoon season covering the following aspects: optimum sowing time, method of sowing, optimum plant population densities and spacing, efficacy of Rhizobial strains, lime requirement in acid soils, manurial requirement, and evaluation of suitable associate crops to be grown with soybeans during the monsoon season in the inter and multiple cropping program of this area. 4) Breeding work has also been intensified to evolve varieties suitable for the winter season to avoid competition with other important crops that can be grown during the monsoon season in similar types of soils. The varieties suitable for growing in both seasons have been established that solve the problem of keeping seeds viable in storage, since the seeds of one season can be used for sowing in the next season.

AGRIBUSINESS

Invited Papers

PROTECTIONISM AND TRADE EXPANSION IN OILSEEDS

J. G. Reed, Continental Grain, New York, New York

Abstract not available at press time.

DEMAND FOR SOYBEAN AND SOYBEAN PRODUCTS IN THE EUROPEAN COMMON MARKET

H. C. Knipscheer, International Institute of Tropical Agriculture, P.M.B. 5320, Ibadan, Nigeria

During the last ten years more than 25 percent of the total U.S. production of soybeans, expressed in soybean meal equivalents, has been exported directly or indirectly to the countries that are presently members of the European Community (E.C.). Variations in West European demand therefore have an important impact on prices of U.S. soybeans, and thus, on the income of U.S. soybean growers and the U.S. trade balance. Economists have been only moderately successful in predicting the changes in the demand for soybean products in the Common Market. This is due in part to the failure to incorporate specific Common Market policy factors into the econometric models that have been used.

The general purpose of this study is to develop an economic model that accurately describes the demand relationships in which the effects of the policies is incorporated. The study is confined to the demand for "aggregate" soybean meal. This is soybean meal in whatever form it may appear on the market, either as soybean meal or still in the form of soybeans. The prediction of future demand for soybean meal and the determination of the effects of the agricultural policies of the European Community on the demand for soybeans and soybean meal within the Common Market are the more specific objectives.

The five goals of the Common Agricultural Policy (CAP) are stated in the Treaty of Rome: 1) to increase the agricultural productivity, 2) to ensure a fair standard of living for the agricultural population, 3) to stabilize the market, 4) to assure supplies to consumers, and 5) to maintain a reasonable price level. The executive branch of the European Community, the Commission, pursues these objectives by enforcing a set of market price regulations. Therefore the prices within the Common Market differ greatly from world prices, in absolute values as well as in relative values. This difference has a large impact on the consumption of meat, the composition of the livestock sector and the use of soybean meal. In the last fifteen years, the consumption of beef decreased relatively while the consumption of pork and poultry meat increased rapidly. With the increases in the hog and the poultry sectors and a series of improvements in feeding technology, the animal feed compound industry grew rapidly, accompanied by similar growth in the demand for protein concentrates. Moreover, soybean meal has an amino acid profile that is very favorable for monogastric animals such as hogs and poultry making soybean meal preferable to other oilseed meals.

Based on the economic principles, the extensive analysis of the effects of the CAP, and experience in the past, the following model was developed. QDSM = f(PSM/PECC, ECPL, APFU, APS and T), where the consecutive symbols stand for quantity soybean meal demanded (QSSM), price soybean meal (PSM), E.C. price cereals (PECC), E.C. profitability index for the livestock sector (ECPL), animal protein feed units (APFU), availability protein substitutes (APS), and a time trend (T). This economic model was tested by the application of three different stochastic models, each based on a different set of assumptions: the Ordinary Least Squares Model, the Error Components Model and the Classical Factor Analysis Regression Model. The economic model performed well in each of the stochastic forms, although interesting differences appeared.

With the help of each of the three statistical models the following effects of changes in the CAP were explored: l) a change in E.C. cereal prices, 2) a change in E.C. meat prices, 3) a change in E.C. skimmed milkpowder production, and 4) a change in E.C. monetary policy was simulated. The policies of the E.C. were shown to have significant effect in shifting the demand for soybean meal. Also, predictions of the demand for soybeans and soybean meal in 1980 and 1985 were compared with estimations that were made by other economists. In general, the results of the study confirmed the steady growth of the demand for soybeans and soybean meal that was previously predicted.

GRADES AND STANDARDS FOR SOYBEANS

L. D. Hill, Department of Agricultural Economics, University of Illinois, Urbana, IL 61801

Grades and standards for grain have three purposes in a market economy: 1) to classify all grain into a few homogeneous categories to facilitate trade, 2) to permit market transactions on the basis of description, and 3) to enable buyers to identify relative value for various end uses. The first two purposes are met by almost any set of descriptive factors, so long as the buyers and sellers accept the standards as the basis for their transactions. The third purpose requires a scientific basis for relating the descriptive characteristics to yield and quality of the products being produced from the grain.

Although soybeans were introduced into the United States around 1912 and U.S. production had increased to 44,378,000 bushels by 1935, official soybean standards were not established until 1940, when an amendment to the U.S. Grain Standards Act of 1917 provided for the inspection and grading of soybeans, requiring official inspection on all soybeans moving in interstate and export trade. The grade factors used were similar to those already in use for other grains—test weight, moisture, total damaged kernels, heat damaged kernels and foreign material. Factor limits for splits and color were included as additional important indicators of quality.

There have been relatively few changes in the grade limits for soybean standards since 1940. They have proven satisfactory as a basis for trade and meet the first two objectives of standards. Recently, however, questions have been raised concerning the ability of the standards to reflect value in processing. There is little evidence that No. 1 soybeans will yield more oil or soybean meal than No. 2 or No. 3. Neither is there a correlation between the quality of oil or meal and the numerical grade of the beans being processed.

The current standards have five deficiencies as a basis for trade in the modern, sophisticated world market. 1) The most important quality factors in terms of end products (protein and oil) are excluded from the grade factors; 2) foreign material includes any material passing through a 8/64" screen, and is generally subtracted as dockage from the weight being purchased even though it contains mostly pieces of soybeans; 3) the grade limits on most factors are arbitrary values; 4) maximum or minimum limits on each factor in the numerical grades encourage blending of beans of different quality levels to include the maximum allowable moisture, foreign material, damaged kernels, etc.; and 5) differences between grades on the factor of heat damage are too small to differentiate using the current sample size of 250 g.

Most proposals for revising the current standards imply a significant cost to marketing firms as a result of changes in procedures, equipment, and grading practices. Proposed changes need to be evaluated to compare the cost with the potential benefits. Grades and standards do not in themselves establish value but only describe relevant characteristics, thus enabling the market to establish value on a uniform basis. Price discounts are the most common technique for differentiating prices among different qualities.

In the United States, price discounts for soybeans are commonly found on most grade factors. These discounts do not remain constant and may change in response to market conditions. Moisture provides an example of these discounts. The level of the discount varies with the price of soybeans, with the moisture levels being harvested, and from one season to the next. Market forces establish this discount level in much the same way as other market forces establish the price level. The weight loss due to moisture removal can be readily calculated. A simple formula determines the discount required to equate the value of two lots of soybeans containing different moisture levels. This formula can be extended to include other grade factors.

Increased soybean production in many countries of the world must be accompanied by the development of a more sophisticated marketing system. In most of these markets a system of grades and standards will be needed to facilitate trade by description. The standards selected should meet the three purposes stated in the first paragraph of this paper. In general, the current U.S. grade standards for soybeans should not be adopted in these new markets without first conducting a careful economic analysis of the alternatives.

ROLE OF SOYBEAN PROTEIN IN NATIONAL DEVELOPMENT

F. H. Schwartz, Ralston-Purina, St. Louis, MO

Abstract not available at press time.

TRENDS IN MARKETING AND DISTRIBUTING SOYBEANS AND PRODUCTS IN BRAZIL

H. Tollini, EMBRAPA, Brasilia, Brasil

Abstract not available at press time.

Contributed Papers

LICENSING NEW SOYBEAN VARIETIES AND TECHNOLOGY

J. L. Helm and S. C. Anand, McNair Seed Company, Laurinburg, NC 28352

License arrangements make it possible for two or more parties to exploit new soybean varieties and technology. Such arrangements allow marketing organizations to circumvent the potential high cost of breeding and development programs. Likewise, certain arrangements can provide breeding programs with increased exposure at minimum risk.

New varieties and technology can be acquired by outright purchase. However, License agreements generally are more attractive. Most of the time, all developmental work has not been completed and actual worth cannot be accurately determined for an outright sale at a fair price. Licensing allows both parties to have a more desirable cash flow and the better understanding of sharing in the profitability of the product. The promotion of new varieties and technology through licensing is a time-consuming and difficult task. Assistance is available from specialized firms whose function is to bring the parties together and promote the products. They share in profitability either on a fee basis or on a proportional basis if their own money was used in advanced development stages.

Protection of all parties' rights is a fact that must be adhered to. We would recommend that secrecy agreements, approved by both parties' legal counsel, be signed and executed before serious discussions begin. If the breeder wants Plant Variety Protection Rights this should be identified early in any discussions. Problems that may arise after an initial agreement should be spelled out in advance of any transfer of material. Problems that require specific statements are:1) rights to improvements, 2) termination provisions, 3) ongoing technical assistance required, 4) market forecasting to be provided and by whom, 5) responsibility for enforcement of breeders rights against infringers, 6) exclusivity provided by each party, and 7) continuation of license if a party is purchased by/or merged into a third party. These and other areas where conflicts could arise will be discussed.

AN ECONOMETRIC MODEL OF U. S. SOYBEAN MEAL AND OIL YIELDS

S. C. Griffin and E. O. Heady, Center for Agricultural and Rural Development, Iowa State University, Ames, IA 50011

Soybean meal and soybean oil typically have been regarded as classic examples of constant fixed-proportion products. Most econometric studies of the soybean industry have assumed soybean product yields to be constant or seasonally exogenous. However, national statistics show considerable seasonal and year to year variation in soybean meal and oil yields.

Economic theory and econometric estimation techniques are combined to better explain the historical variation in U. S. soybean product yields. Statistically significant variables include soybean meal and oil prices, soybean stocks and flows at mills, percent of crushing capacity utilized by the industry, and rainfall during specific periods of the soybean growing season. The yields of soybean meal and oil were found to be complementary, other things equal, rather than competitive (i.e., the higher the meal yield, the higher the oil yield). The price elasticities of product yields were also found to be very low.

U.S. SOYBEAN INDUSTRY AND ITS ROLE IN WORLD MARKETS

S. A. Gazelle, Fibers and Oils Program Area, USDA, CED/ESCS, Washington, DC 20250

Soybean production in the United States increased from a small, relatively obscure crop of around 5 million bushels in the mid-1920's to its current size of nearly 2 billion bushels. It is now the world's major oilseed commodity. From that humble beginning, soybeans became a leading cash crop for American farmers and also rank at the top of the list as a major U. S. agricultural export commodity.

The dramatic expansion in United States soybean acreage and production was fueled primarily by burgeoning domestic requirements for soybean meal, required to feed expanding livestock and poultry numbers which mushroomed after World War II. Concurrently, technological and marketing improvements in the production and consumption of margarine, shortening, and cooking and salad oils opened a vast outlet for soybean oil. As these sophisticated food industries became adopted overseas, global demand for soybean products expanded dramatically.

Today, United States soybeans account for nearly one-half of total major world oilseed production. Or put another way, the oil and meal derived from U.S. soybeans now supply over a tenth and about two-fifths, respectively, of total world production of all fats and oils and of all major high-protein meals. They account for nearly a fifth and roughly one-half, respectively, of the total volume of these commodities moving in world trade. Approximately two-thirds of the world's soybeans are produced in the U.S. We account for over three-fourths of the world trade. In recent years, the U.S. share has declined, due to increasing soybean production in other parts of the world.

A little over one-half of our total soybean crop (including soybean equivalent of products), is exported. During the 1977/78 marketing year, the equivalent of nearly 26 million metric tons of soybeans was exported, or roughly 54 percent of the 1977 crop of 48 million metric tons. Total dollar value for these exports amounted to nearly $6.5 billion, the top agricultural export commodity and a significant contribution to our balance of payments posture.

During 1977/78, the United States exported the equivalent of nearly 21 million metric tons of soybean meal (5.5 million as soybean meal and about 15 million in the soybeans that were exported). Soybean oil exports during the same period totaled nearly 4.5 million metric tons (0.9 million as oil and 3.5 million in the soybeans).

With world demand for soybeans still expanding, the U.S. again this marketing season is supplying a large share of global requirements. A little over 50 percent of the 1978 U.S. soybean crop (including bean equivalent of products), is expected to be exported.

PRICE AND POLICY ANALYSIS USING A STRUCTURAL MODEL OF THE SOYBEAN AND SOYBEAN PRODUCT MARKETS

W. H. Meyers and D. Hacklander, Commodity Economics Division, USDA, ESCS, Washington, DC 20250

An econometric model of the soybean industry is used to analyze the impact of key factors in domestic and foreign markets. The objective is to determine the sensitivity of soybean and soybean product markets to selected external variables.

A simultaneous supply and demand model of the soybean industry is solved given a base set of exogenous data. Each of the key exogenous variables is changed one at a time and a new solution is obtained. The difference between the equilibrium levels in the new solution and those in the base solution provides a vector of impact multipliers for the endogenous variables.

One of the key domestic factors is the price of corn, which affects both current demand and plantings for the following year. A $.10/bu. increase in the farm price of corn would increase the current year farm price of soybeans by $.04/bu. The less favorable price ratio with corn would reduce soybean supply the following year by 27 million bushels. This would lower crush, exports and ending stocks by 8, 6 and 13 million bushels, respectively. The supply reduction raises the farm price of soybeans $.18 bu. above the base level. The decline in crushing reduces supplies of meal and oil and causes these prices to rise by $6/short ton and .6 cents/lb., respectively.

Two key export market variables are the strength of the dollar relative to foreign currencies and Brazil's soybean exports. A 10 percent decline in the value of the dollar increases soybean exports by 30 million bushels, 15 of which comes out of stocks and 15 out of crush. The farm price of beans increases by $.19/bu. and the prices of meal and oil increase by $2.3/ton and 1.2 cents/lb., respectively. A one million m.t. increase in Brazil's soybean exports would decrease U.S. exports by 29 million bushels, 18 of which are added to stocks and 11 to domestic crush. The prices of beans, meal and oil decline by $.22/bu., $7.9/ton and .7 cents/lb., respectively.

The impact multipliers generated by the model capture the simulteneous interactions of the bean, meal and oil markets and provide a convenient method for tracing the affects of key external factors on the equilibrium levels.

OIL AND PROTEIN DETERMINATIONS—AN OBJECTIVE APPROACH TO U. S. SOY-BEAN STANDARDS

P. E. Parker and F. F. Niernberger, Standardization Division, USDA, FGIS, Kansas City, MO 64105

Inspectors presently assign numerical grades to soybean samples using objective and subjective evaluations of quality factors shown in the official United States Standards for grain. Most of the grading factors currently employed in soybean grades do not appear to exhibit a high degree of correlation with oil or protein content, the two major measures of end-use quality. Oil and protein factor determinations have not been employed in the federal grain standards due to difficulties in rapid measurement and field inspection conditions. Recent advances in near-infrared reflectance (NIR) and nuclear magnetic resonance (NMR) technologies now permit rapid objective measurement of soybean protein and oil content. Collaborative studies are currently in progress to determine uniform sample preparation of soybeans, reproducibility of intra and inter-laboratory samples, and standardized instrument operating procedures.

Implementation of oil and protein testing for soybeans could be done on a voluntary or requested inspection basis with the results recorded on the grade certificate for informational purposes. This approach would allow adequate testing under actual marketing practices and conditions prior to any mandatory requirements. A similar approach was adopted for protein determinations of Hard Red Spring and Hard Red Winter Wheats using the NIR by the Federal Grain Inspection Service in May of last year. The advantages of using precise objective grading methods for soybean inspection focus on improving classification accuracy and reproducibility. The addition of oil and protein determination on the inspection certificate is centered on more closely identifying end-use quality. Growers are encouraged to produce high oil and protein soybeans through the market pricing mechanism and buyers benefit from the knowledge that the grade better characterizes the quality of the soybeans for their intended end-use.

PHYSICAL PERFORMANCE AND ECONOMIC FACTORS ON OVERSEAS SHIPMENTS OF SOYBEANS

C. J. Nicholas and A. Abdul-Baki, USDA, SEA/FR, Beltsville, MD

In 22 soybean test shipments originating in the United States and unloaded overseas, both foreign material (FM) and splits increased during movement from origin to destination, the former from an average of 1.6 to 1.8 percent, and splits from 12.2 to 14.2 percent. Soybeans are handled 15 to 20 times while moving from the farm to the overseas receiver. Breakage or damage increases in direct proportion to the amount of handling. The number of handlings not only increases the damage but also increases the cost. Industry sources estimate marketing costs increased as much as 2 cents per bushel per handling. A significant analysis of the data is the development of the FM, splits, and fines data. Of the eight test shipments analyzed, one-third of the samples had fines which made up one-half or more of the FM content. All of the shipments had fines constituting more than one-third of the FM. Fines in FM indicate the amount of soybeans damaged or broken. Shortage of grain hopper cars and barges is a most serious transport problem confronting the soybean shipping. Another transport problem is the high transport costs for moving the soybeans from the U.S. interior terminal elevator to the port of embarkation.

Soybeans are augured, conveyed, dumped, or otherwise handled at least 15 times between the farm and the overseas customer. Of the 12 overseas shipments analyzed, fines averaged 0.96 percent. Computerizing this 0.96 average on the annual soybean exports, a dollar loss figure of $42 million is projected. In addition to this loss, soybeans are also subject to neutral oil loss and losses due to splits, FM, which all amount to serious dollar loss due to breakings in transport and handling. Insect infestation continues to present a problem, especially to receivers in the Far East. Seven of the 12 test shipments destined for Japan were insect infested and required fumigation. Analysis of weights between "invoice" and "loaded weight" in 12 of the test shipments showed weight shortages on six shipments varying from 0.4 to 0.8 percent and averaging 0.5 percent. Analyses of four Brazilian soybean shipments showed the oil higher than in U.S. shipments by about 1 percent, protein content about the same, and free-fatty acids higher in the Brazilian beans, although the neutral oil losses were about the same as the U.S. beans.

SOYBEAN PRODUCTION IN AUSTRALIA

L. C. Campbell, P. W. G. Sale and O. G. Carter, Department of Agronomy and Horticultural Science, University of Sydney, N.S.W. Australia 2006

Attempts to establish soybeans as a commercial crop in Australia prior to the later 1960's were not successful. Disappointing and unreliable yields were produced as farmers tried to grow American cultivars under Australian conditions. Following integrated research programs on the East Coast of Australia during the late 1960's, successful farming techniques were established together with a knowledge of which cultivars to grow in the different growing areas. The national average yield of 770 kg/ha (11.3 bu/acre) in 1967-68 has doubled to 1516 kg/ha (22.5 bu/acre) in 1977-78.

The unique feature of the Australian soybean industry is that at least 80 percent of soybeans are grown under irrigation or supplemental irrigation. At present the crop is quite popular among irrigation farmers in New South Wales and Queensland. Its competitive position among other summer crops together with significant but limited increases in available irrigation water in certain districts, suggest an increase in the production of irrigated beans. The dryland soybean areas along the coast of New South Wales and Queensland and adjacent tableland regions also offer scope for increased production. New techniques for the direct drilling of soybeans into the acid, coastal soils show promise. As this enterprise offers an attractive alternative to beef raising which is the present form of land use, it is felt that soybean farmers will be attracted to these areas.

In 1977-78 it is estimated that over 73,000 tons were produced in Australia. The present demand for soybean products in Australia is for approximately 80,000 tons of meal and 48,000 tons of oil. A crop of 100,000 tons would supply the nation's meal requirement, with the deficiency of oil being supplied by imports. Any additional production would tend to be exported as whole beans, as facilities for exporting meal are limited.

It is anticipated that Australia's domestic requirement for soybean meal will be provided by locally grown beans within the hext few years. The capacity to greatly expand soybean production, as has occurred in Brazil, is not present in Australia owing to the unreliability and lack of suitable summer rain over most of the continent, and to the lack of irrigation resources.

MARKETING, TRANSPORT AND STORAGE

Invited Papers

PRINCIPAL DETERMINANTS OF VARIATIONS IN SOYBEAN PRICES

R. M. Lamm, USDA, CED/ESCS, Washington, DC 20250

Over the last decade, soybean prices have varied from a low of $2.30 per bushel at Decatur in 1969, to a high of $10.84 per bushel in June 1973. Soybean oil prices have ranged from 7.3 cents per pound in October 1968 to a high of 43.3 cents per pound for Decatur crude in August 1974, while soybean meal prices peaked at $412.50 in June 1973 following a low of $69.80 in January 1969 on a per ton basis for Decatur 44 percent protein. Nominal price variations of this degree are significant when compared with those of nonagricultural products, and complicate the planning process for both consumers and producers.

The United States has experienced varying inflation rates over the last decade, however, and significant nominal price variation would be expected for all commodities as a consequence. Transforming nominal soybean and soybean product prices to constant 1972 dollars reduces the price variability of these commodities considerably, but not sufficiently to negate the detrimental effects of price instability on planning.

On an international basis, the market for soybeans is virtually a free market. No major importing country imposes tariffs on soybean or soybean meal imports, although many countries impose ad valorem tariffs on soybean oil. In this respect, the U.S. price of soybeans is essentially the world price for soybeans since the U.S. is the largest producer and exporter of soybean and soybean products. For this reason, events in other parts of the world can often have significant influence on U.S. soybean and soybean product prices.

Variations in soybean prices are caused by changes in demand for soybean oil and meal, as well as by changes in soybean supply. Regarding oil and meal, there have been five factors contributing to increased demand over the last decade. These include: 1) increased real incomes in both developed and developing countries, 2) changes in consumer tastes in favor of more meat consumption and a demonstrated preference for vegetable oil over highly saturated animal fats, 3) higher relative prices for oils and meals which can be substituted for soybean oil and meal in consumption, 4) world population growth, and 5) higher inflation rates which stimulate the holding of commodities as a hedge.

With respect to the supply of soybeans, there have been four factors contributing to price variability. These include: 1)technology advances, which have been widely diffused, 2) changes in the costs of inputs used in soybean production, 3) weather variation, predominantly droughts in various parts of the world, and 4) changes in the prices of alternative products which soybean producers can grow as an alternative to soybeans. Drought in some areas of the world, leading to decreased supply of oilseeds and oilseed substitutes, has been the principal cause of price variation over the last decade, although other factors also have contributed.

SOY OIL UTILIZATION–CURRENT SITUATION AND POTENTIAL

O. R. Erickson and R. A. Falb, American Soybean Association, St. Louis, MO

Soy oil dominated the United States edible fats and oils market with a commanding 60 percent of the total edible fats and oils market and as high as 80 percent of the margarine market and 90 percent of the prepared dressings market. On a world-wide basis in 1978, soy oil accounted for approximately 23 percent of world production while its nearest competitor, sunflowerseed, accounted for only 9 percent. However, in spite of this market dominance, research conducted by the American Soybean Association shows there is little general knowledge of soy oil. This research will be reported. In a study using unidentified samples of competitive commercially available oils, soy oil compared very favorably in home conditions in all aspects. This research will be reported.

The stability of soy oil can be improved by special processing where economic circumstances dictate. Although extensively done in the U.S. this presently is not always feasible in other parts of the world. Nutritionally soy oil is regarded as equivalent to competitive premium vegetable oils as an unhydrogenated oil, but additional testing must be and is being done on lightly hydrogenated oil of commerce.

Some problems still have to be solved, such as the linolenic content, which in many countries limit the potential because of stability, food regulations, costs, etc. A solution to some of these problems could greatly increase the potential. In many countries, oil utilization is the key to the increased potential market for soybeans. A country which cannot use the oil has the alternatives of exporting the oil or importing soybean meal. Neither of these alternatives are normally as economically attractive as processing soybeans and utilizing both the oil and the meal.

RISK MANAGEMENT IN MARKETING SOYBEANS

D. E. Kenyon, Department of Agricultural Economics, Virginia Polytechnic Institute and State University, Blacksburg, VA 24061

Price risks in marketing soybeans have grown substantially in the 70's compared to the 60's. Potential for improving profits through better management is greater in the area of pricing than production. Producers have essentially three major pricing alternatives—cash, cash contracting, and hedging via futures. Pricing strategies for each alternative are compared in terms of average price and price variance. Pricing strategies considered are based on: 1) cost of production, 2) distribution of sales over production and marketing season, 3) technical indicators of price direction, and 4) basis patterns during storage. Results indicate that a significant tradeoff exists between average price and price variance.

COMPETITIVE POSITION OF OILSEED SUNFLOWERS WITH SOYBEANS

H. O. Doty, Jr., Fibers and Oils Program Area, USDA, CED/ESCS, Washington, DC 20250

Sunflowers compete with soybeans for acreage and for markets in end products. This competition has increased recently and is expected to become more intense in the years ahead. Competition for acreage will grow due to the yields of hybrid oilseed sunflowers increasing relative to soybean yields. As farmers learn how to handle the sunflower crop and plant improved varieties (higher oil content and better disease and insect resistance) yields will increase.

Oilseed sunflowers have been commercially produced in the United States for the past twelve years primarily on land formerly planted to flax, wheat, and cotton. It is no longer a curiosity, an experimental crop, or a fad. Production and yields have increased in most of the intervening years since its introduction. In 1977, a large-scale switch from open pollinated varieties to hybrids occurred, yields increased 25 percent. According to plant breeders sunflowers are in the same stage of development that corn was in the early 1930s. Over 3 million acres were planted to sunflowers this year. This is more acreage than is individually devoted to such crops as peanuts, flaxseed, rye, and rice. In the near future expansion of sunflower is likely on the fringe of present corn, soybean, and cotton growing areas and land formerly planted to wheat or barley.

Sunflowers are an oilseed crop grown primarily for its oil. Roughly 75 percent of sunflower's value is obtained from the oil and only 25 percent from its high protein meal. In contrast, soybeans are an oilseed crop grown primarily for its high protein meal. In recent years roughly 60 percent of the value of soybeans has come from the high protein meal and 40 percent from the oil. On a per pound basis the oil is more valuable than the meal. Oilseed sunflowers are over 40 percent oil whereas soybeans are about 18 percent oil.

Competition between sunflowers and soybeans will also take place as their oils compete for markets in oil products. There is now a dependable supply of sunflowers for crushing in the U.S. and consumers have indicated a strong demand for sunflower oil products. In comparison with soybean oil sunflower oil requires less refining, has a higher smoke point, and is considerably higher in polyunsaturation. These characteristics make it particularly desirable for use in making high quality margarine and cooking and salad oils. The primary advantage that soybean oil has over sunflower oil is its lower price. This gives soybean oil the edge in some major fat product markets such as vegetable shortening and low priced vegetable oil margarine. Because of the world protein shortage it is anticipated there will be relatively little competition between the high protein meals of soybeans and sunflowers.

EXPORTING SOYBEANS THROUGH THE PORT OF NEW ORLEANS

W. M. Gauthier, G. R. Hadder, and H. D. Traylor, Department of Agricultural Economics and Agribusiness, Louisiana Agricultural Experiment Station, Louisiana State University, Baton Rouge, LA

United States soybean exports in 1977 exceeded 595 million bushels. Over 90 percent of exports passed through the Port of New Orleans. In 1977, over 1.5 billion bushels of grain, approximately 44 percent of all U.S. grain exports, were exported through the Port of New Orleans. The potential of the New Orleans grain export system to handle soybeans and other grains differs from its performance because of bottlenecks. Within this dynamic system, bottlenecks can be caused by any possible combination of technical, economic, and/or institutional sets of variables.

The grain export system is a configuration of stationary export elevators and mobile modes that perform complementary and competing transport, storage, and handling functions. Reproduction of the performance of that system as measured by the volumes of grain passed through each elevator during 1977 provides the basis for analyzing the potential of the system to handle the projected 1985 volume. In addition, it provides the framework for analyzing other scenarios of interest for a particular elevator, set of elevators, or the system.

A bottleneck is any impairment of grain arrival and/or departure. The index of performance is the annual volume of grain throughput. Conceptually, average lengths of barge, rail, and ship queues measure quantities of grain in mobile storage that would be in transit if not for the existence of some bottleneck(s). This measure ignores the economic incentive for storage in mobile modes. The severity and incidence of a bottleneck is given by the maximum length of the barge, rail, or ship queue at an elevator. The quantity measurements of bottlenecks are complemented by relative measures of utilization at specialized modal unloading or loading facilities. Simultaneous executions of the model for successively larger volumes for a particular elevator or for the system identify the most likely points of bottleneck(s) for each elevator in the system. The incorporation of demurrage costs provides for an economic assessment of the costs of bottlenecks within the system.

SOYBEAN PROCESSING INDUSTRY—U.S. AND FOREIGN

J. J. Mogush, Cargill Inc., Minneapolis, MN 55440

Soybeans today are processed for meal and oil on all of the continents of the world with the exception of Antarctica. The growth of the industry has occurred because of strong demand by consumers for finished products, the production of which requires vegetable protein meals and vegetable oil. Soybeans have met the needs of the world consumer better than any other oilseed. Currently, world soybean production is 54 percent of total world oilseed production. The soybean is the only oilseed which has increased its market share in the past ten years.

Geographic location of soybean processing plants in the world has been determined largely by economic factors. The two most important of these are: 1) a plentiful supply of soybeans, and 2) a ready market for soybean meal. Other important factors are a ready market for oil, a reliable and competitive transportation system, a competent and competitive supply of labor, and a plentiful supply of competitively priced energy.

Restrictive international trade practices have had and will continue to have an influence on processing plant locations and expansion. The Brazilian export subsidy program encourages expansion in Brazil and discourages expansion in other countries which might otherwise be well placed economically to serve world product markets. The EEC has an important duty structure which favors its own processing plants. The industry is structured in fragments which encourage vigorous competition. Private single plant owners compete with multi plant owners, multi-national, co-ops, and government owners. Neither sector is likely to take over the industry.

Efforts to improve technology have focused on reducing energy and making labor more productive. Significant progress has been achieved in conserving energy. We have seen process improvements achieved by combining several pieces of equipment into one such as the meal desolventizer dryer and coller. More waste heat is being recovered. More heat exchangers are being used.

The trends of the past thirty years strongly suggest continued growth for the world soybean processing industry. Increasing population and rising standards of living will require larger supplies of high protein meal and edible oils. Soybean meal and soybean oil have proved to be excellent products. Substantial increases in production of soybeans will be required. It will have to come through either higher yields or more acres or both. The U. S., no doubt, will be a major contributor to increased world production of soybeans. A proportional increase in U. S. soybean processing capacity will not occur. The U. S. already has very high utilization rates for soybean meal and soybean oil. Increased demand for these products is more likely to occur in other countries. Economic and political realities are likely to favor the location of most of any expanded processing capacity in foreign countries.

WEED CONTROL

Invited Papers

ECOLOGY OF SOYBEAN/WEED INTERACTIONS

C. G. McWhorter and D. T. Patterson, USDA, SEA-AR, Southern Weed Science Laboratory, Stoneville, Mississippi 38776

The evolutionary responses of weeds in soybean production relate mainly to biological features that enable certain species to cause interference with the growth and reproduction of the soybean plant. Interference is the combined effect of competition plus allelopathy. Competition is defined as a simultaneous demand by more than one organism for the same resources when the immediate supply of the resources is below the combined demand of the organisms. Allelopathy is defined as the detrimental effect of one plant on another caused by the synthesis and release of toxic or inhibitory substances. Characteristics that contribute to the competitiveness of weeds are: a) life-form, b) vegetative reproduction, c) similarity to the crop; d) food reserve capability, e) efficient and rapid uptake of water and nutrients, f) rapid partitioning of dry weight into leaf area, g) efficient uptake of CO_2, h) efficient use of water, i) rapid rate of growth, and j) phenotypic plasticity.

Of the 37 worst weeds in soybeans in the United States, 14 are monocotyledons and 23 are dicotyledons. All of the 13 monocotyledons are C_4 plants, while 22 of the 23 dicotyledon plants are C_3 plants. Ten of the 14 monocotyledons are exotic, while 12 of 23 dicotyledons are exotic. Allelopathy has been demonstrated in 8 of the 14 monocotyledons and 12 of the 23 dicotyledons. Allelopathy will likely be demonstrated in several others in the future.

New weed problems often arise as the result of: a) genetic changes, b) chemical selection, c) distribution, and d) cultural selection. New weeds arising as a result of genetic changes are apparently rare, but this has been poorly investigated. Changes in weed problems in soybeans due to chemical selection, distribution and cultural selection have been commonplace in the United States. Several examples of changes in weed problems that have occurred in the past are discussed along with changes now being experienced by soybean producers.

WEED CONTROL SYSTEMS IN THE CORN BELT STATES

F. W. Slife, University of Illinois, Department of Agronomy, Urbana, Illinois 61801

Soybean acreage has grown rapidly in the Corn Belt States in the last 10 years and in some areas it is approaching equal status acreage with corn. Improved weed control technology is at least partially responsible for the increased acreage.

In the large cash grain areas of the Corn Belt, soybeans are usually rotated with corn. This two crop system has tremendous advantages in so far as weed control is concerned. A wide choice of weed management practices are available for corn and when used properly, this reduces the weed problem in the soybean crop that follows.

In the southern third of the Corn Belt, winter wheat is frequently included in the crop rotation. Most of the wheat is double cropped to soybeans using zero tillage techniques. Weed problems are usually more severe in this system and it requires extra weed management effort.

The major weed problem in soybeans in the Corn Belt is summer annual weeds that have the same life cycle as corn and soybeans. The most prominent species is giant foxtail *(Setaria faberi)*. *Digitaria* and *Panicum* spp. are also troublesome. Some 8 to 10 annual broadleaf species infest the soybean acreage with varying intensity. This latter group of weeds is the major concern of soybean growers because with the present management practices, they are not consistently controlled.

Perennial weeds are present throughout the Corn Belt but with the exception of johnsongrass *(Sorghum halepense)*. They are of less concern to soybean growers than the complex of annual weeds.

Weed management practices most widely used in soybeans at the present time are: 1) tillage before plant to destroy emerged weeds; 2) application of a grass herbicide either preplant or preemergence; 3) an herbicide for broadleaf weed control is usually combined with the grass herbicide treatment, but it may be applied postemergence; 4) a shallow tillage (rotary hoe or similar tool) is often used just prior to soybean emergence or shortly after to control those annuals that escape the herbicide treatment; 5) row cultivation (1, 2, or 3 times) is used as the need arises; and, 6) some hand weeding is done on some of the acreage to improve the appearance of the fields and to prevent reinfestation of weed seeds.

Weed control techniques are in a constant state of change in the Corn Belt. More and more growers are developing a growing awareness of their particular weed problem through pest management programs that emphasize the use of scouts.

Reduced tillage systems are not always compatable with some of the standard weed control practices and, hence, the weed control program has to be changed.

Weed control in soybeans has reaced a high degree of sophistication in the Corn Belt and soybeans are now grown in fields that previously were considered too weedy to grow the crop. Although most of the soybean crop is grown in rows of 76 centimeters or wider, the trend toward very narrow rows in increasing. This has been made possible through good weed management practices.

WEED CONTROL SYSTEMS IN SOUTHERN U.S.

Robert Frans, Department of Agronomy, University of Arkansas, Fayetteville, Arkansas

The Southern U.S. is a region of high temperatures and rainfall, both conducive to luxuriant crop and weed growth. The competitive effect of weeds with soybean growth necessitates optimum levels of control. Generally, soybean varieties grown are of the bushy type, lapping rows relatively early, and giving shading of soil. Weed infestations prior to current herbicide usage were a broad range of annuals, both grass and broadleaf. Now broadleaf weeds that are difficult to control predominate, as well as johsongrass.

Systems of weed control employed are somewhat dependent upon weeds infesting the fields. Tillage coupled with incorporation of dinitroaniline herbicides is utilized for grass weed control. The same herbicides are often applied at double rates for rhizome johnsongrass control as well as dalapon and glyphosate. The latter two herbicides are applied to foliage, a practice that often delays soybean planting unduly. Acetanilides, such as alachlor and metolachlor, or triazines, such as metribuzin, are applied preemergence or at planting, for general control of annuals. At soybean emergence, contact herbicides such as dinoseb or naptalam plus dinoseb are applied for control of emerging cocklebur or morningglory seedlings.

Early postemergence herbicides are applied at soybean V2 to V4 and may include dinoseb or bentazon overtop. Post-directed applications also start at these growth stages and may include chloroxuron, dinoseb, or naptalam plus dinoseb. A little later, or around V4 to V6, linuron plus 2,4-DB may be applied post-directed for control of newly-emerging vining weeds. Most of these practices may be repeated several days later, particularly if weeds persist. Considerable interest has been generated in the South for late-season control of johnsongrass emerged above the soybean canopy. Applications are accomplished with innovative recovery or rope wick applicators with such herbicides as glyphosate. A new herbicide family, the diphenylethers, offers much promise for selective control of the grass weeds, and mixtures with selective broadleaf herbicides may further decrease post-directed practices in current use.

Soybeans are grown full-season in standard rows, utilizing various of the above practices according to need. Reduced or minimum tillage soybeans utilize a preemergence herbicide plus a contact material applied at planting in previously undisturbed seedbeds. This sytem is useful on rolling lands where wind or water erosion may be a problem. Soybeans are often planted following winter small grains, either directly into the grain stubble, or in stubble burned and disked. Conventional or narrow row widths are used. Full-season, narrow-row bean culture is increasing, particularly with the advent of better weed control. There is little yield advantage compared to conventional width rows, but considerable savings in fuel and labor costs. The presence of perennial weeds often limits adoption, since there is no cultivation option available with such a practice. Red rice, a particular problem in areas of soybean-rice rotations, is attacked in the soybean crop through use of preplant incorporated herbicides, such as double rates of dinitroanilines or alachlor, followed by post-directed applications of paraquat in conventional row-width soybeans.

Contributed Papers

PENNSYLVANIA SMARTWEED AND COMMON RAGWEED INTERFERENCE IN SOYBEANS

H. D. Coble, North Carolina State University, Raleigh, North Carolina 27650

Currently, representatives of the pest disciplines are taking an important step toward truly integrated pest management, i.e., adjusting philosophies and approaches in using the general concept of economic thresholds. If integrated pest management is to become a functional part of a total soybean crop management system, well-defined economic thresholds must be developed for specific pests and pest complexes. The first step in establishment of economic thresholds for weeds is the completion of field interference studies to determine the damage threshold for the different weed species. These studies were undertaken to determine the damage thresholds for Pennsylvania smartweed *(Polygonum pensylvanicum L.)* and common ragweed *(Ambrosia artemisiifolia L.)* separately in soybeans.

Weed density treatments consisting of 0, 2, 4, 8, 16, or 32 equi-distantly spaced weeds per 10 m of row were used to determine the threshold density of a full-season infestation. In order to determine the critical weed-free period, two types of treatments were used: a) a natural weed population was allowed to grow for 0, 2, 4, 6, or 8 weeks after crop emergence, followed by weed-free maintenance for the remainder of the growing season; and b) weed-free maintenance for a period of 0, 2, 4, 6, or 8 weeks after crop emergence followed by natural weed reinfestation.

Linear regression analysis showed the damage threshold for Pennsylvania smartweed in soybeans to be approximately three weeds per 10 m of row, whereas for common ragweed, the damage threshold was nearly eight weeds per 10 m of row.

No yield reduction resulted from either species if the crop was kept weed-free for 4 weeks or longer. Likewise, no yield reduction resulted from either species if weeds were removed during the first 6 weeks of the growing season.

POSTEMERGENCE CONTROL OF ANNUAL GRASSY WEEDS IN SOYBEANS WITH HOELON[R] (HOE-23408)

J. A. Grande, E. T. Palm, and P. W. Robinson, American Hoechst Corp., Somerville, New Jersey

In 1977 and 1978, HOE 23408 was tested extensively as a postemergent herbicide for the control of annual grassy weeds in soybeans. Experimental use permit trials were designed to determine the efficacy of the herbicide when applied over a wide range of environmental conditions using various application techniques. Small plot research trials compared efficacy of the herbicide using formulations with and without an adjuvant. Plots ranged in size from 0.001 acres for research trials to 20 acres for experimental permits. Spray volumes of 20-40 gallons were tested. Flat fan, cone, or flood jet nozzles were used in permit trials. Use rates of 0.75, 1.00, and 1.25 lb active ingredient/acre were tested. Applications were made when the majority of annual grassy weeds had 1-4 leaves present.

HOE 23408 effectively controlled most annual grasses when applications were made at the proper stage of growth, 1-4 leaves. Later applications were less effective. All rates tested gave excellent annual grass control, with the higher rates providing more consistency. Differences between spray volume did not significantly effect control. Differences between nozzles did cause variability in control indicating that good spray coverage was important. The adjuvant formulation did not consistently improve grass control at the optimum use rate. Phytoxicity increased with the adjuvant formulation at application, but was not evident later in the season. Soybean yields with both formulations were superior or equal to standard soybean herbicides.

Excellent postemergent control of annual grasses in soybeans can be obtained with HOE 23408. HOE 23408 use in soybeans will complete the postemergent weed control spectrum. Confidence in postemergent weed control should encourage an increasing number of soybean producers to plant narrow row or solid seeded stands of soybeans.

POSTEMERGENCE CONTROL OF VOLUNTEER CORN IN SOYBEANS WITH HOELON^R (HOE-23408)

J. A. Grande, E. T. Palm and P. W. Robinson, American Hoechst Corporation, Somerville, New Jersey

Hoelon^R (HOE-23408) was extensively evaluated for the postemergence control of volunteer corn in soybeans across the midwestern and northeastern United States during 1978. The studies were carried out in both small plot trials and under an Environmental Use Permit Program with selected farmers. The variables under study in this program included volunteer corn stage of growth, use rate of Hoelon, and susceptibility among volunteer corn of different genetic backgrounds.

Results indicate that Hoelon is very effective for the postemergence control of volunteer corn. Spray timing should coincide with the stage of growth of the volunteer corn. The optimum timing of Hoelon applications is considered to be when the majority of volunteer corn is between 8" and 12" tall. Earlier applications may not allow for the majority of volunteer corn plants to emerge. The use rate of Hoelon ranges from 0.75 to 1.0 pound of active ingredient per acre, depending on the infestation level of volunteer corn. There appears to be some genetic influence on volunteer corn susceptibility to Hoelon. This, however, does not appear to be a serious problem. More information on plant morphology as affected by genetic background needs to be developed. Most volunteer corn plants emerge from an ear. This creates a situation when as many as thirty plants are emerging from a clump. Thorough spray coverage is essential in this situation with the spray boom positioned above the volunteer corn plants. Soybean tolerance to postemergence Hoelon application is excellent.

EVALUATION OF HERBICIDAL ANTIDOTES FOR WEED CONTROL IN SOYBEANS

A. S. Bhagsari, Fort Valley State College, Fort Valley, Georgia 31030

The control of weeds in soybeans, especially broad leaf weeds, is a costly multistep operation involving preplant incorporation, preemergence and postemergence herbicide applications and cultivations. Metribuzin, an asymmetrical triazine is an effective preemergence herbicide for the control of many broad leaf weeds in soybeans. Use of metribuzin in soils of low organic matter gives erratic weed control and also results in considerable crop damage when heavy rains follow shortly after herbicide application. Use of herbicide antidotes, called protectants or safeners, seem to enhance the crop tolerance limits to herbicides, with no such effect on most weed species. Increased selectivity through antidotes can permit higher use levels of metribuzin than the recommended one's for effective weed control in soybeans. This practice could eliminate the need for postemergence herbicide applications and/or cultivations. The objective of this study was to evaluate if certain potential antidotes will protect soybeans against metribuzin toxicity resulting from higher than recommended rates.

The study was conducted in a greenhouse as well as in the field using Bragg and CNS cultivars of soybeans. In the greenhouse, potential antidotes, like dexon, phthalimide, and others, were applied as seed coating (0.05 and 0.1 percent of seed weight) and were also incorporated in the soil in the concentration range of 25 to 400 ppm. The plants were harvested after 35 days. During 1977, a preliminary field experiment was conducted to determine the optimum dose of metribuzin when applied preemergence. In 1978 a preplant dose of 0.30 lvs/A of metribuzin was applied followed by preemergence application of an additional 0.30 and 0.60 lbs/A. Immediately after planting, activated charcoal was applied at 150.0 and 300.0 lbs/A, in a 4" wide band over the row. before preemergence application of metribuzin. A hand weeded check was included. Three supplemental irrigations, about 2.0" of water each time, were given during 1978. No irrigation was applied in 1977.

No injury to soybeans was observed due to metribuzin applied preemergence at 0.30 to 0.90 lbs/A probably due to low rainfall in 1977 and 1978. There was no significant difference in the yield of soybeans in 1977. Weed population estimates in control plots after 20 days of planting the soybeans mainly showed the presence of pigweed, grasses and coffeeweed. Preplant incorporation of metribuzin at 0.30 lbs/A failed to control grasses and coffeeweed. But the preplant dose combined with 0.30 and 0.60 lbs/A of metribuzin gave satisfactory control of most weeds. Application of metribuzin two to three times higher than the recommended rates had no effect upon plant height and number of branches per plant. Due to weed competition the dry matter per plant at pod formation stage was reduced by 20 to 30 percent where metribuzin was applied at 0.30 lbs/A. Nitrogenase activity for both the cultivars was the lowest in the plots treated with 0.90 lbs/A of metribuzin. The incorporation of 25 to 400 ppm phthalimide (as an antidote) in soil containing 0.2 ppm metribuzin seemed to give less protection to the cultivar Bragg than CNS.

BIOLOGICAL CONTROL OF WEEDS WITH FUNGAL PLANT PATHOGENS—WITH EMPHASIS ON COCKLEBUR IN NORTH CAROLINA

C. Gerald Van Dyke, North Carolina State University, Raleigh, North Carolina 27650

Weeds cause an estimate 10 percent loss each year in the production and quality of food, feed, and fiber crops in the United States of America. About $2.7 billion is spent yearly for cultural, ecological, and biological methods of weed control. Thus, the annual losses caused by weeds and the cost of their control amount to about $11 billion. Prevailing conditions and attitudes are likely to dictate an increased emphasis on integrated pest management systems for future weed control programs, including the use of biocontrol agents.

The objectives of the Southern Regional Project are: 1) To identify and determine the distribution of fungi infecting cocklebur in North Carolina (NC); and 2) To evaluate the potential of fungal pathogens as biocontrol agents for reducing weed problems.

NC counties were continually surveyed during the 1978 growing season to identify and isolate fungal pathogens on cocklebur. Pathogens were collected for future study. Some were cultured when possible or propagated on cocklebur plants. Herbarium mounts were made for future references to plants and their collection sites. Detailed studies of pathogens isolated were done with light and scanning electron microscopy (SEM). Greenhouse inoculations of cocklebur were done to determine initial reactions of the weed to pathogens.

Soybean and other crops were surveyed for fungus infected cocklebur in 1978. A rust fungus (*puccinia xanthii* Schw.) has been consistently collected from cocklebur throughout the state. Some leaves were heavily infected with the rest. Light and SEM studies of this fungus, which is microcyclic and autoecious, revealed the germination of teleospores and infection processes. Cocklebur plants inoculated in the greenhouse show symptoms in about one week after inoculation. These studies are being done to determine optimum conditions for obtaining infection, and the extent of damage to cocklebur with varying concentrations of inoculum. Preliminary results of studies of this rust in Australia and in NC suggest it has potential as a biocontrol of cocklebur.

FLOURESCENT DYE SHOWS INCORPORATION PATTERNS OF EQUIPMENT

L. Thompson, Jr., W. A. Skroch and E. O. Beasley, North Carolina State University, Raleigh, North Carolina 27650

Uniform distribution and precise placement are essential for optimum efficacy and efficiency of herbicides incorporated into the soil. Herbicides mixed in soil cannot be seen, but in ultraviolet light (UV) flourescent dyes can be seen in soil at night. Therefore, a fluorescent dye was used to determine incorporation patterns with various kinds of equipment. The dye was sprayed on the soil surface with a hand sprayer. Incorporation equipment passed over the sprayed area. A shallow pit was dug, exposing a vertical face (profile) of soil approximately 20 cm deep and 2 to 3 m wide. Using UV light and time-lapse photography at night, the results were observed and recorded. The power-driven vertical tillers provided perfect incorporation to the exact depth of operation. A single pass with a tandem finishing disc at 4 mph caused streaking, a repeated pattern of heavy deposition of the dye and no dye, corresponding to distribution of the blades on the disc. After bedding a "single-disced" area, the streaks were still very prominent. Cross discing (twice at 90°) resulted in a pattern similar to the power-driven vertical tiller, but less uniform and more variable in depth. When a "cross-disced" area was bedded, there was a deep uniform distribution of dye in the bed. By knocking-off much of the bed (as at planting), a uniform distribution of the dye in the bed and in the middles was achieved. When the dye was sprayed into the rear of the spinning (stirring) tines of a power-driven horizontal tiller, a very uniform, shallow (3.8 to 5.0 cm) pattern was observed. Dye patterns produced by several field cultivators were less uniform than those mentioned above. The distribution of dye applied to bedded land and incorporated with a rolling cultivator depended on the condition and size of the bed and on the way the rolling cultivator was set. Power-driven vertical tillers with variable length or uniform length times provided excellent incorporation into beds.

FIRST YEAR OF STUDIES ON THE PHYTOTOXIC RESIDUALITY OF SOIL-APPLIED HERBICIDES IN URUGUAY

A. Fischer, Nitrasoil, Florida 622, P. 4, 1005 Buenos Airies, Argentina, and A. Tasistro, Gelati 58 Dept. G, Mexico 18, DF Mexico

The possible residual phytotoxicity on subsequent crops by several soil-applied herbicides, some of which are known as having a significant residuality, was evaluated. Several reports point out the phytotoxic effect of the Cl-triazines and trifluralin several months after application. These products are being used in this country mainly with sorghum, maize and soybean crops.

Trifluralin, presowing incorporated (PSI), 0.96 k a.i/ha; metribuzin, preemergence (PRE), 0.50 k a.i/ha; prometryne, (PRE), 2.67 k a.i/ha, were applied on soybean between December 2-6, 1976. Atrazine 2.4 k a.i/ha was applied on sorghum, (PSI), and incorporated with tooth harrow on November 11, 1976. Five months after treatment, random samples were taken from the first 10 cm of treated soil (brunizem; pH 5.6; organic matter 3.3 percent; and 23.3 meq/100 g Cation Exchange Capacity). The samples were potted in a complete randomized design, corresponding 16 pots to each product where oats, soybean, lucerne and flax were sown in 4 replications. The trial was laid in the greenhouse. Thirty days after sowing, the aerial part of the plants was harvested and dry weight was assessed. A second lot of soil samples was taken 40 days after the first extraction. The same procedure as above was followed.

Significant growth reductions were observed, in part due to heavy rainfalls on the treated area, which were far above normal records and might have had a marked leaching effect. Nevertheless, lucerne showed a higher susceptibility to trifluralin residues, as well as flax to atrazine and oats to metribuzin. Soybean appeared to be the least affected species in all cases. Alachlor had no inhibitory effects.

Considering the observed (although insignificant) tendencies, the possible effect of the mentioned rainfalls, and the bibliography reviewed, it is still possible to expect adverse effects on certain subsequent crops with the use of residual soil herbicides, particularly with higher doses and dry weather.

WEED CONTROL AND MANAGEMENT OF HERBICIDES IN DOUBLE-CROPPED SOYBEANS IN BRAZIL

G. Davis, USAID, EMBRAPA, University of Wisconsin/Tennesse Tech. University, E. Voll, CNP Soja, EMBRAPA, Londrina, Parana, Brazil, and H. Lorenzi, IAPAR, Londrina, Parana, Brazil

Most of the soybeans in Brazil are planted in a double crop sequence with wheat. Both crops are produced on a commercial scale. Herbicides are applied to about 60 percent of the soybeans. The crop is planted in row spacings of 50 to 60 cm on land with slopes as great as 15 percent. Inter-row cultivation is difficult. Minimum tillage is practiced on about 5 percent of the soybean hectarage.

The major soybean production zone in Brazil is located between 20 and 30 degrees S. latitude. Two distinct production regions (northern and southern) are recognized. The regions are distinguished by the length of the time period between the harvest date of wheat and the optimum planting date for soybeans. In the northern region, wheat is harvested six weeks before the optimum date for soybean planting. In the southern region, what harvest and soybean planting dates coincide. Certain weedy grasses will produce 800 kg/ha of dry matter in the six weeks between wheat harvest and soybean planting.

Weed infestations in the older soybean fields of Brazil are extremely high. The predominent grassy weeds are species of *Brachiaria* and *Digitaria*. In eight field trials in the State of Parana in 1975-76 and 1976-77, the dry weight of *Brachiaria* in the weedy control plot averaged 3500 kg/ha. The most prevalent broadleaf weeds in soybeans are species of *Bidens* and *Sida*. Weed species composition in soybean fields changed rapidly in response to herbicide pressure applied to prevalent species. Infestations of *Bidens* followed control of grasses with trifluralin. Broadleaf species tolerant to linuron and metribuzin invaded plots to which these herbicides were applied for two years.

The maximum rate of herbicide that soybeans will tolerate is required to control weeds in the crop in Brazil. Carryover in the soil of herbicide in amounts equal to 20 percent of the rate applied to soybeans is toxic to wheat seedlings. Data from three years of research indicate that carryover levels of soybean herbicides in Brazil are rarely toxic to wheat in the double-crop system.

RESEARCH TECHNIQUES

Contributed Papers

APPLICATION OF VON BERTALANFFY'S FUNCTION TO SOYBEAN GROWTH MODELING

R. Guarisma, Central University of Venezuela, Maracay, Venezuela

The application of an extended form of von Bertalanffy's function to soybean [*Glycine max* (L.) Merr.] hypocotyl elongation data is considered. Hypocotyl elongation of 'Forest 1975' soybean cultivar was recorded for 10 consecutive days at 48 combinations of soil temperature, soil moisture, and soil mechanical impedance. A combination of multiple regression models and von Bertalanffy's function should have more flexibility than the autocatalytic, monomolecular or Gompertz functions as a mathematical model for predicting average soybean hypocotyl elongation under stress conditions of the soil physical environment. The information contained in von Bertalanffy's function is conveyed very conveniently and its constants yield information of direct biological interest.

SIMULATION OF SOIL COMPACTION UNDER TRACTOR WHEELS

H. D. Bowen, North Carolina State University, Raleigh, North Carolina and H. Jaafari, Arya-Mehr University of Technology, Isfahan, Iran

Twenty years ago an approach to predicting soil compaction under tractor wheels was reported in the literature. The approach used Boussenesq's equation with Froehlich's concentration factor to predict the major principal stress σ_1. It was assumed that the soil compaction was determined by σ_1 alone and by substitution into a confined compression equation, the compaction in terms of bulk density could be determined. Boussinesq's equation for determining principal stress has only partially been confirmed, but no other approach to soil compaction prediction has been offered. It was the objective of this study to evaluate the reported approach using 30.5 cm soil cores taken from the center of the tractor tire track, sliced into 2.5 cm sections, and bulk density determined by gravimetric means.

Three tractors ranging in size from 1724 kg to 5904 kg were operated on freshly plowed field with no drawbar load and moisture at near field capacity. Boussinesq's equation in conjunction with confined compression tests on a Wagram sand and an Appling sandy clay loam predicted too high a bulk density from the tire-soil interface to a depth of half the tire width below it. Bulk density directly below center of the track at depths greater than half the tire wide compared favorably with measured bulk densities.

A zone of maximum compaction was found at half the tire width below the tire-soil interface where the major principal stress was much less than its maximum, but where a zone of maximum shear stress τ_{max} and shear strain γ_{max} occurred. The fact that the maximum compaction occurred at a location where the major principal stress is not at its highest value proves conclusively that the major principal stress σ_1 does not uniquely determine soil compaction.

In summary, Boussinesq's equation may not be in serious error for predicting the major principal stress σ_1 on a uniformly tilled soil, but the assumption that σ_1 when substituted into confined compression test data will accurately predict compaction is seriously challenged where shear strain is not negligible. Multiplier coefficients were developed for the two soils so that the computer model would predict compaction directly under the center of the track for no draft loading and moisture near field capacity.

INFORMATION SOURCES FOR SOYBEAN RESEARCH

P. J. Boyle and R. J. Lewington, Commonwealth Bureau of Pastures and Field Crops, Hurley, Maidenhead, Berks SL6 5RL, United Kingdom

After an introductory survey of the development of published research on soybeans, the range of aspects needing to be covered by information services and the main forms in which the primary literature on soybeans is published are surveyed and estimates of the number of literature records in each of the main subject areas given. This is followed by accounts of the main secondary information sources and services, including the development of on-line services.

A MANUAL SEEDER FOR SOYBEAN

G. Singh, N. Thunyaprasart and P. A. Cowell, Asian Institute of Technology, Bangkok, Thailand

Soybean is frequently sown in paddy fields following rice harvest without any land preparation in Thailand, Taiwan and other Asian countries. It is sown by making on opening (about 2 cm deep) close to rice stubble with a hand trowel. Two to four seeds are dropped in each opening. Most farmers do not cover the seeds. Generally female labor is employed. Female laborers are required to either bend over or squat while sowing which is very tiresome.

The Asian Institute of Technology has developed a simple seeder. It consists of a long handle, a sliding valve, a soil opener, a sliding tube and a metering device. The handle serves as seed tube and carries about 2 kg of seeds. The metering device is roller-type and has two seed cells. The size of seed cells is adjusted to seed size. A crank mechanism is connected to the sliding tube. Every time the sliding tube moves up and down the roller rotates and seeds drop in the opening. In testing here at the AIT and at the AVRDC (Asian Vegetable Research Development Center, Taiwan) the seeder looks very promising. Up to 20 percent of the time can be saved in seeding using this seeder compared to manual seeding in addition to reducing the drudgery.

ADDENDUM

PEST PROBLEMS OF SOYBEANS AND CONTROL IN NIGERIA

M. I. Ezueh and S. O. Dina, National Cereals Research Institute, Moor Plantation, Ibadan, Nigeria

Soybeans, *Glycine max* (L.) Merrill were first introduced in Nigeria in 1910 (Mayo, 1945). Production is limited to about 170,000 ha of land mostly in the Savannah zone with an annual output of 70,000 metric tons. Cultivation is usually in mixtures with sorghum, millet and citrus orchards. In the traditional management of the crop, no crop protection measures are applied and the average yield is low, about 300 kg/ha (Edem, 1975).

As part of an effort on soybean improvement which began at the National Cereals Research Institute (NCRI) in 1974, a survey of the insect fauna of the crop was carried out between 1974 and 1977 to identify insects of economic importance in the production of the crop as a sequel towards planning protection measures. Preliminary insecticide trials were undertaken in two seasons.

Pest complexes of soybeans closely resemble those of cowpea in Nigeria. Although the economic significance of these has not been fully ascertained, this will increase as production scale rises. Insect damage, use of poor varieties and poor management practices are responsible for the present low yields in plots. Preliminary insecticide trials have not shown much promise yet probably because of poor understanding of the biology of the pest complex in relation to the plant host. Three applications may be inadequate for season-long protection.

INTERCROPPING SOYBEANS WITH CEREALS IN EAST AFRICA (TANZANIA)

R. K. Jana, Department of Crop Science and Production, University of Dar Es Salaam, P.O. Box 643, Morogoro, Tanzania

Intercropping sometimes referred to as mixed cropping is the cropping system that involves the growing of two or more crops in the presence of one another on the same piece of land. It is dominant in peasant farming in the tropics and also in East Africa it is a traditional cropping practice. However, a legume is normally included in the intercrop with cereals. The major legumes in Tanzania are groundnuts, beans, cowpeas and pigeon peas.

Soybeans are a recently introduced crop into Tanzania peasant farming and are expected to occupy a prominent position as a food stuff/export crop in the near future. Hence, it is important to see whether soybeans are suited to the existing intercropping system so as to popularize them among the peasant farmers in Tanzania.

In this study at Morogoro (Lat. 6°S, altitude 525 m), the crops that were chosen for experimentation included two legumes, soybean and green gram, and three cereals, maize, sorghum, and millet. Each legume was intercropped with either of the cereals with alternate rows, alternating within a row, or paired rows of legumes. The grain yield data of three years' average indicated that soybeans grew well in combination with all the cereals tested. Further, soybeans gave a relatively higher yield when planted in alternate rows with cereals. The results suggest that soybeans have a place in the intercropping system of the peasant farmers of Africa, with particular reference to Tanzania.

GENETIC VARIABILITY INDUCED IN SOYBEANS BY FAST NEUTRONS AND BY N-NITROSO-N-ETHYL UREA (NEU)

S. Muszynski and A. Dabrowska, Institute of Genetics and Plant Breeding, Warsaw Agricultural University, ul. Nowoursynowska 166; 02-766 Warsaw, Poland

Marked progress was achieved in mutation breeding of soybeans, and several valuable mutants as well as even varieties were obtained, e.g. as in Japan, USSR, Thailand or Democratic Germany. In most cases gamma rays were applied to induce mutations in soybeans. In the field experiments described here, the effects of NEU and of n_f were compared in several cultivars of soybeans. It appeared, however, that soaking the soybean seeds in water solutions of NEU or even in distilled water caused a considerable decrease in seed germination, whereas irradiations did not produce such an effect. The highest dose of n_f, 2000 rads, diminished the field germination to the 50% level of controls. In the M_2 generation, few chlorophyll mutants as well as one climbing mutant were found, and an increase in genetic variability of quantitative traits was noticed.

CYTOGENETIC EFFECTS OF FAST NEUTRON IRRADIATION OF SOYBEAN SEEDS AS MEASURED BY LEAF SPOTTING

S. Muszynski and A. Dabrowska, Institute of Genetics and Plant Breeding, Warsaw Agricultural University, Ul. Nowoursynowska 166; 02-766 Warsaw, Poland; and J. Huczkowski, Institute of Nuclear Physics, ul. Radzikowskiego 152, 31-342, Cracow, Poland

Dry seeds of two soybean cultivars, Fiskeby and R11/17, were irradiated with n_f in a cyclotron of the Institute of Nuclear Physics, Cracow, and the dosages applied were 400, 800, 1200, 1600 and 2000 rads. The same two cultivars as well as four others were given the n_f treatment in the impulse reactor of the United Institute of Nuclear Research in Dubna, USSR, with the dosages of 240, 700, 1100 rads. The seedlings obtained from the irradiated seeds showed numerous chloratic spots on their first leaves, whereas controls did not. Statistical analysis of the data showed that the relationship between the effect and the dosages applied was exponential. It was also shown that the cultivars reacted differently, thus demonstrating the genetic differences in radiosensitivity between cultivars. The two kinds of n_f applied produced different cytogenetic effects as measured by the chloratic-spot test. It is suggested that the number of chloratic spots on soybean leaves can be used as a very convenient and sensitive test of cytogenetic effects of ionizing radiations.

SCREENING, ISOLATION AND ZONAL TRIALS OF SOYBEAN LINES SELECTED FROM THE AVRDC ADVANCED GENERATIONS

L. Rahman, M. S. Hoque, M. B. H. Sikder, and M. Ali, Bangladesh Agricultural University, Mymensingh, Bangladesh

Seeds of nineteen F_4 and five F_5 pedigree populations of soybean were introduced from The Asian Vegetable Research and Development Center (AVRDC) in the year 1975. These breeding lines were sown in the Bangladesh Agricultural University Farm, Mymensingh with a view to isolate the better ones. The isolated lines were tested in subsequent generations both in "Kharif" (April-September) and "Rabi" (October-March) seasons to know their adaptibility and to select still better types. Of all these materials, 33 lines have been developed—25 from the original F_4 populations and 8 from the original F_5 populations. These selected lines were put under observation trials in three test centers. Growth, yield, and other characteristics have been studied.

RESPONSE OF SOYBEANS ON SANDY SOILS TO IRRIGATION STRATEGIES

L. C. Hammond, J. W. Jones, and K. J. Boote, University of Florida, Gainesville, Florida

Shallow and frequent irrigation is being proposed as a water and nutrient-conserving technique for sandy soils in humid climates. Our objective was to evaluate various water management strategies in terms of water-use efficiency, water depletion from the soil profile, and growth and yield of soybeans. In 1978, soybeans were grown on a deep, well-drained Lake fine sand under five water management treatments (overhead sprinkler). The treatments were: (1) control, no irrigation, (2) irrigation to a soil depth of 22 cm when suction at 15 cm was 200 to 500 millibars, (3) irrigation to a soil depth of 45 cm scheduled the same as (2), (4) irrigation to a soil depth of 45 cm when the suction at 30 cm was 200 millibars (after a 10-day delay in the start of irrigation), and (5) irrigation depth and schedule were variable—same as (2) except irrigation to a soil depth of 45 cm when the suction at 30 cm was 200 millibars.

Rainfall supplied an overall excess of water until the beginning pod stage (R_3). There was 3.3 cm of rainfall from this point to maturity (75 days). The respective treatment irrigation amounts (cm) and numbers for the 75-day period were: 0, 17.6 (19), 20.9 (17), 11.9 (10), and 18.0 (16). Seed yields were 980, 2600, 2800, 2300, and 2700 kg/ha, respectively. The relationship of yield and water applied during the 75-day period was:

$$y = 1053 + 90x;\ R^2 = 0.86$$

The results show a soybean yield response to total water applied rather than to irrigation strategy. Yields in treatment four were reduced by the 10-day stress and perhaps by the less frequent irrigation schedule. The irrigation strategies were not designed for the unusual seasonal and the long-term drought experienced. However, the data indicate that shallow, frequent irrigation was as effective as deeper, less frequent irrigation. Thus, the water and nutrient savings benefits from the former strategy should be realized under normal short-term drought conditions.

SOME RESULTS FROM THE USE OF EMS TO INDUCE MUTATIONS IN VARIETIES OF SOYBEANS, *GLYCINE MAX*

C. A. Panton, Department of Botany, University of West Indies, Jamaica; and E. P. Montaque, Ministry of Agriculture, Hope, Jamaica, West Indies

Studies carried out for three generations on three varieties of soybean plants raised from seeds soaked for 6 hr in 0.05 M E.M.S. showed that in the first generation, damage to embryo cells resulted in 37% reduction in survival, 42% in growth rate, 15% stem fasciation and forking, and a high incidence of malformed leaves and stems. Chlorophyll mutants appeared in the second generation in frequencies ranging from 0 to 15%. M_3 selections yielded putative mutants for early flowering, high yield, improved standing ability, increase in number of leaflets per leaf, changes in seed coat color and up to 5% in seed protein. It is suggested that by inducing greater genetic diversity in soybeans, E.M.S. can be used successfully for exploiting heritable variation in the crop.

PROSPECTS FOR SOYBEANS IN SOUTH EAST ASIA

T. B. Thiam, University of Malaya, Kuala Lumpur, Malaysia

In 1977, South East Asia produced only about 1% of the total world production of soybeans. The harvested area of soybeans is less than 800,000 ha. Imports of soybeans into these countries, however, are quite considerable and total more than $100 million. Soybeans are imported in the form of whole grain and oil for human consumption and in the form of cake and meal for animal consumption. The bulk of soybean import is in the form of whole grain, cake, and meal.

The import of soybean oil is relatively insignificant. South East Asia is the major producer of lauric and palm oil. Malaysia is the leading palm oil producer and exporter in the world. Indonesia is second. Production of lauric oils from copra and palm kernel meal are mainly from the Philippines, Indonesia and Malaysia. These countries are responsible for over 80% and 60% of the world trade in palm oil and lauric oil. It is envisaged that production of palm oil will continue to increase rapidly in the next few years and by 1980 palm oil will capture approximately 27% of the world trade in vegetable oils and will rank second only to soybean oil.

South East Asian countries have in recent years shown an increasing interest in cultivating soybeans. This interest stems from the rapid increase in demand for soybeans and soybean products. Soybeans are being viewed as an attractive crop to satisfy the increasing demand for food arising from both population and income growth. An intensive research program is being undertaken currently in several of these countries to find suitable varieties of soybeans that can give high yields under local climatic and agronomic conditions. For the immediate future, South East Asia will remain a soybean deficient region and will continue to import increasing amounts of this commodity from the major soybean producing counties.

SOYBEAN RESEARCH AND PRODUCTION IN ZAMBIA

F. Javaheri, P.O. Box 11, Magoye, Zambia

Experiments were conducted between 1968 and 1978 to obtain recommendations for soybean production under Zambian conditions. Large numbers of varieties were tested and some were released for commercial production. In 1979, an estimated area of 2500 ha will be grown with an expected average yield of 1800 to 2000 kg/ha. Hale 3 (early) and Davis (medium) adapted to top management conditions, Bossier and Geduld (medium) adapted to all conditions and Hermon 147 (late) with natural nodulation adapted better to unfavorable conditions. Other varieties under seed production which may be released are Santa Rose and UFV1 (Brazilian) and a selection from 71-38 a breeding line from Dr. Byth of Queensland.

A breeding program has been initiated with emphasis on tolerance to shattering and later flowering under our conditions. Two years' experimental work has indicated a possibility of two (and in some cases three) generations per year in a breeding program. Two efficient strains of Rhizobium have been selected and a strong soybean response of inoculation has been obtained. Evaluating the residual effect of nitrogen fixation has indicated that soybeans fixed an average of 170 kg N/ha per year of which 86 kg was contributed to the subsequent crop. Additional research is needed on nodulation, minimum tillage, harvest losses, germination and problems of soybean production by small scale farmers.

PACKAGES OF TECHNOLOGY FOR SOYBEAN PRODUCTION IN BANGLADESH: AN AGRO-ECONOMIC ANALYSIS

M. Zahidul Hoque, Bangladesh Rice Research Institute, P.O. Box 911, Dacca, Bangladesh

Soybeans are a new crop in Bangladesh. Attempts were made during the early forties and late fifties to introduce the crop in the country but without any success. The main reasons for failure in fitting the crop in the cropping systems of Bangladesh were the unavailability of suitable varieties, crop damage in the rainy season, lack of irrigation facilities in the dry season, and poor farmer response. The poor farmer response might have been due to lack of marketing facilities, low demand for soybean oil, and farmers' unfamiliarity with the cultivation method.

In today's world, soybeans have been recognized as one of the cheap but rich sources of protein. Considering the wide-spread protein malnutrition in Bangladesh, emphasis is now being given to the introduction and extension of soybeans at the farm level. By this time, soybean oil has also become familiar and commands a high market price.

A national coordinated soybean research project with the Bangladesh Agricultural Research Council as the coordinating agency was started in 1975. To date, the project has generated a wealth of useful information for production and utilization of soybeans in Bangladesh. The agro-climatic conditions of Bangladesh are quite suitable for soybean cultivation. It could have an immediate potential production area of about 120,000 ha with a production potential of about 240,000 tons of soybeans.

INSECT PESTS OF SOYBEANS IN TROPICAL AFRICA

S. R. Singh, International Institute of Tropical Agriculture, Ibadan, Nigeria

Soybeans, *Glycine max* (L.) Merrill, as an agricultural crop in tropical Africa, are usually cultivated at national government research stations and on government-supervised farms as the sole crop. In addition some soybeans are grown by farmers in Nigeria and Kenya. One of the major limiting factors for soybean cultivation in tropical Africa has been seed germination. There is a big potential and need for cultivation of this crop as a source of valuable human, animal, and poultry feed, and oil for the industry.

Due to their high susceptibility to pests, the traditional grain legumes grown in tropical Africa are often considered major-risk crops. At present, this is not the situation with soybeans. There are comparatively few economically important pests; those pests that are present can be controlled effectively by insecticides. However, for any realistic increase in soybean production at subsistence-farm levels, pest control will soon be essential. A sound integrated pest management system has to be developed. However, the necessary information on the components of such a pest management system is lacking. Therefore, there is urgent need to study the various components of a pest management system involving the survey and bionomics of the pests and biological agents, various relatively undisturbed traditional farming systems, and their effects on soybean insects. Pest-resistant varieties undoubtedly will play a major role in the pest management system and this could form a valuable component of the crop improvement programs. Resistance to some of the minor pests like leafhoppers should be easy to incorporate, and as a matter of policy all developed or commercial varieties should have resistance to this pest. In addition, these varieties should not be highly susceptible to either a minor or major pest. The aim should be that the future soybean varieties continue to be minimally dependent on insecticide applications.

GENOTYPE X ENVIRONMENT INTERACTIONS IN THE SOYBEAN BREEDING PROGRAM OF RWANDA DURING THE LAST DECADE

P. Nyabyenda and M. J. J. Janssens, Institut des Sciences Agronomiques du Rwanda, Rubona, BP 138, Butare, Rwanda

Nine multienvironmental trials have been conducted in two ecologically different places, Rubona and Karama, as part of the soybean research program of Rwanda during the last decade. The GXE interaction effects were generally highly significant. Trials were characterized by small genetic coefficients of variation, small estimates of broad-sense heritability and very limited expectations of genetic advance. Combining the experimental results of trials conducted separated at Rubona and Karama enable one to conclude that each region required specific genotypes. However, two genotypes, Davis and Palmetto, with general environmental adaptability would be identified. Davis was found to be significantly superior to Palmetto.

By plotting the GXE interaction effects and by performing a cluster analysis, groups of genotypes with similar productivity and environmental adaptability could be isolated. These groups did not correspond with the genetic and geographic origins of the genotypes nor with the maturity groups. Differences of yields were related to localities rather than to seasons. The future breeding program will aim at generating a greater genetic variability by means of massive introductions and crosses. Multienvironmental testing will be enhanced with more emphasis placed on localities.

AUTHOR INDEX

www.ingramcontent.com/pod-product-compliance
Ingram Content Group UK Ltd.
Pitfield, Milton Keynes, MK11 3LW, UK
UKHW022201190926
475702UK00003B/136
* 9 7 8 0 3 6 7 2 1 4 0 4 3 *